For my Oma, Gerda Lukas
1911–2006
May she rest in peace.

And for my children,
Guthrie and Harper.
May they live in peace.

elin o'Hara slavick

BOMB AFTER BOMB

A VIOLENT CARTOGRAPHY

foreword by Howard Zinn
essay by Carol Mavor
interview by Catherine Lutz

For Marty + Ellen,
May we know a better world.
In gratitude + with peace + hope.
elin o'Hara slavick
2009

CHARTA

Design |
Gabriele Nason

Editorial Coordination |
Filomena Moscatelli

Editing |
Emily Ligniti

Copywriting and Press Office |
Silvia Palombi Arte&Mostre, Milano

Sales Department |
Antonia De Besi

US Office |
Francesca Sorace

Cover |
World Map, Protesting Cartography:
Places the United States Has Bombed, 1854–Ongoing

Photo Credits |
The Mint Museum of Art: Alamogordo, New Mexico, USA, and Vieques Island, Puerto Rico, USA
Christopher Ciccone and Karen Malinofski for the North Carolina Museum of Art: World Map; Congo; El Salvador; Shifa Pharmaceutical Plant, Sudan; Haiti; Lebanon; Amchitka Island, Alaska, USA; Afghanistan I; Korea
Christopher Ciccone: all other drawings

We apologize if, due to reasons wholly beyond our control, some of the photo sources have not been listed.

ISBN-10: 88-8158-633-9
ISBN-13: 978-88-8158-633-2

Printed in Italy

Edizioni Charta srl

via della Moscova, 27 - 20121 Milano
Tel. +39-026598098/026598200
Fax +39-026598577
e-mail: edcharta@tin.it

US Office
New York City, Tribeca
Tel. +1-313-406-8468
e-mail: international@chartaartbooks.it

www.chartaartbooks.it

Acknowledgements |
I thank the curators and institutions that have exhibited this work, including: Linda Johnson Dougherty at the North Carolina Museum of Art in Raleigh; Jose Marti National Library in Havana; the Annex in New York; Kathy Hudson at the John Hope Franklin Center at Duke University, Durham, North Carolina; Gerard Brown for the Borofsky Gallery in Philadelphia; Arti et Amicitiae in Amsterdam; Studio 84 and 5+5 Gallery in Brooklyn; the Ashmore Gallery in Miami; Gallery 110 in Seattle; Rx Gallery in San Francisco; Amy Cappellazzo and Laura Hoptman for the Weatherspoon Art Museum in Greensboro; and Todd Smith and Carla Hanzal at the Mint Museum in Charlotte, North Carolina.

I thank the University of North Carolina at Chapel Hill for research support, especially the Institute for Arts and Humanities. Thanks also to the North Carolina Arts Council for a generous Fellowship. I especially thank the writers who were willing to contribute an essay to this book once upon a time: Christian Appy, John Armitage, John Berger, Robert Celestial, Ariel Dorfman, Tom Engelhardt, Cynthia Enloe, Stan Goff, Sven Lindqvist, Rania Masri, John Pickles, Peter Turchi, Terry Tempest Williams, and Denis Wood. Alas, this is a different book.

I am indebted to many people for making this project and book possible: Jane Marsching for her initial encouragement; William Blum, Sven Lindqvist, and Howard Zinn for their radical and truthful scholarship and Howard Zinn again for his generous foreword; Catherine Lutz for showing me the Homefront, for organizing teach-ins with me, and for the interview for this book; Carol Mavor for her unfolding essay, for being the best friend, teacher, art historian, writer, and mother in the world; Sarah Miller for helping Carol Mavor with her essay; Rob Seguin and Susanne Slavick for editorial assistance; my sisters Susanne and Sarah Slavick for their dedicated examples; my sister Madeleine Slavick for her poetry, faith, and support; my parents Ursula and William Slavick for making me political at birth, for introducing me to art, struggle, and education, and for their incredible activism; Andrew Johnson for absurd energy; Joy Garnett for early editorial suggestions; Amy Ruth Buchanan for good and generous advice; Joel Sternfeld for continuing inspiration; Lisa Ross for being there; Joel Brouwer for sustained solidarity, friendship and poetry; Val Martinez and Laura Guinan for friendship; Laura Sharp Wilson for honesty; Taj Forer for introducing me to Charta Books, for *Daylight Magazine*, and for being an unforgettable graduate student and photographer; Giuseppe Liverani and Francesca Sorace at Charta Books for their belief in this work; my children Guthrie and Harper Ursula who tell me to find it when I lose my temper; and most of all my comrade David Richardson for solidarity and love.

Contents

Foreword

Howard Zinn

Perhaps it is fitting that elin o'Hara slavick's extraordinary evocation of bombings by the United States government be preceded by some words from a bombardier who flew bombing missions for the U.S. Air Corps in World War II. At least one of her drawings is based on a bombing I participated in near the very end of the war—the destruction of the French seaside resort of Royan, on the Atlantic coast.

As I look at her drawings, I become painfully aware of how ignorant I was, when I dropped those bombs on France and on cities in Germany, Hungary, Czechoslovakia, of the effects of those bombings on human beings. Not because she shows us bloody corpses, amputated limbs, skin shredded by napalm. She does not do that. But her drawings, in ways that I cannot comprehend, compel me to envision such scenes.

I am stunned by the thought that we, the "civilized" nations, have bombed cities and countrysides and islands for a hundred years. Yet, here in the United States, which is responsible for most of that, the public, as was true of me, does not understand—I mean really understand—what bombs do to people. That failure of imagination, I believe, is critical to explaining why we still have wars, why we accept bombing as a common accompaniment to our foreign policies, without horror or disgust.

We in this country, unlike people in Europe or Japan or Africa or the Middle East, or the Caribbean, have not had the experience of being bombed. That is why, when the Twin Towers in New York exploded on September 11, there was such shock and disbelief. This turned quickly, under the impact of government propaganda, into a callous approval of bombing Afghanistan, and a failure to see that the corpses of Afghans were the counterparts of those in Manhattan.

We might think that at least those individuals in the U.S. Air Force who dropped bombs on civilian populations were aware of what terror they were inflicting, but as one of those I can testify that this is not so. Bombing from five miles high, I and my fellow crew members could not see what was happening on the ground. We could not hear screams or see blood, could not see torn bodies, crushed limbs. Is it any wonder we see fliers going out on mission after mission, apparently unmoved by thoughts of what they have wrought?

It was not until after the war, when I read John Hersey's interviews with Japanese survivors of Hiroshima, who described what they had endured, that I became aware, in excruciating detail, of what my bombs had done. I then looked further. I learned of the firebombing of Tokyo in March of 1945, in which perhaps 100,000 people died. I learned about the bombing of Dresden, and the creation of a firestorm that cost the lives of 80,000 to 100,000 residents of that city. I learned of the bombing of Hamburg and Frankfurt and other cities in Europe.

We know now that perhaps 600,000 civilians—men, women, and children—died in the bombings of Europe. And an equal number died in the bombings of Japan. What could possibly justify such carnage? Winning the war against Fascism? Yes, we "won." But what did we win? Was it a new world? Had we done away with Fascism in the world, with racism, with militarism, with hunger and disease? Despite the noble words of the United Nations charter about ending "the scourge of war," had we done away with war?

As horrifying as the loss of life was, the acceptance of justifications for the killing of innocent people continued after World War II. The United States bombed Korea, with at least a million civilian deaths, and then Vietnam, Cambodia, Laos, with another million or two million lives taken. "Communism" was the justification. But what did those millions of victims know of "communism" or "capitalism" or any of the abstractions that cover up mass murder?

We have had enough experience, with the Nuremberg trials of the Nazi leaders, with the bombings carried out by the Allies, with the torture stories coming out of Iraq, to know that ordinary people with ordinary consciences will allow their instincts for decency to be overcome by the compulsion to obey authority. It is time, therefore, to educate the coming generation in disobedience to authority, to help them understand that institutions like governments and corporations are cold to anything but self-interest, that the interests of powerful entities run counter to the interests of most people.

This clash of interest between governments and citizens is camouflaged by phrases that pretend that everyone in the nation has a common interest, and so wars are waged and bombs dropped for "national security," "national defense," and "national interest."

Patriotism is defined as obedience to government, obscuring the difference between the government and the people. Thus, soldiers are led to believe that "we are fighting for our country" when in fact they are fighting for the government—an artificial entity different from the people of the country—and indeed are following policies dangerous to its own people.

My own reflections on my experiences as a bombardier, and my research on the wars of the United States, have led me to certain conclusions about war and the dropping of bombs that accompany modern warfare.

One: The means of waging war (demolition bombs, cluster bombs, white phosphorus, nuclear weapons, napalm) have become so horrendous in their effects on human beings that no political end—however laudable, the existence of no enemy—however vicious, can justify war;

Two: The horrors of the means are certain, the achievement of the ends always uncertain;

Three: When you bomb a country ruled by a tyrant, you kill the victims of the tyrant;

Four: War poisons the soul of everyone who engages in it, so that the most ordinary of people become capable of terrible acts;

Five: Since the ratio of civilian deaths to military deaths in war has risen sharply with each subsequent war of the past century (10% civilian deaths in World War I, 50% in World War II, 70% in Vietnam, 80–90% in Afghanistan and Iraq) and since a significant percentage of these civilians are children, then war is inevitably a war against children;

Six: We cannot claim that there is a moral distinction between a government that bombs and kills innocent people and a terrorist organization that does the same. The argument is made that deaths in the first case are accidental, while in the second case they are deliberate. However, it does not matter that the pilot dropping the bombs does not "intend" to kill innocent people—that he does so is inevitable, for it is the nature of bombing to be indiscriminate. Even if the bombing equipment is so sophisticated that the pilot can target a house, a vehicle, there is never certainty about who is in the house or who is in the vehicle;

Seven: War, and the bombing that accompanies war, are the ultimate terrorism, for governments can command means of destruction on a far greater scale than any terrorist group.

These considerations lead me to conclude that if we care about human life, about justice, about the equal right of all children to exist, we must, in defiance of whatever we are told by those in authority, pledge ourselves to oppose all wars.

If the drawings of elin o'Hara slavick and the words that accompany them cause us to think about war, perhaps in ways we never did before, they will have made a powerful contribution towards a peaceful world.

Blossoming Bombs

Carol Mavor

I am remembering the film from which this voice of summer was frozen inside a marble and now emerges.

> At seventeen seconds after 8:15, on the clear bright morning of August 6, 1945, an atomic bomb was dropped on Hiroshima, Japan. The pilots of the United States Air Force's 509th composite group could see flowers in the gardens below . . .
> Two hours later, drops of black rain the size of marbles began to fall.

These words are from the voice-over, a man's voice, sad, deliberate, slow, and desperately urgent, from a little, explosive documentary film that I saw in high school: *Hiroshima-Nagasaki, August 1945*, shot right after the blast by Akira Iwasaki and produced in 1970 by Erik Barnouw.[1] Amazingly, Barnouw was able to get his hands on top secret footage that had been held and suppressed for almost a quarter of a century by the U.S. Department of Defense.[2] Not only can I still hear the clear, male American voice-over that speaks the incomprehensible facts of Hiroshima in a poetics that moves one with all of the simplicity of the silent blast of a Haiku poem, I can also hear the voice of a Japanese girl, as she gives her personal account of the bombing. The words are those of an unnamed Hiroshima girl who had survived on the edges of the blast. She speaks only once, after the start of the film, after the facts are given. Hers is a beautiful and battered voice. She reads in broken English.[3] Hers is the voice of a flower miraculously growing in poisoned soil. This girl's voice haunts this unforgettable film and now, especially, haunts me. I can only hope that it will haunt you. She speaks:

> I remember, I remember, a big light comes, a very strong light, I never see so strong.
> I did not know what has happened.
> My friend, she and I are always together, but I could not find her.
> So dark it gets. So red like a fire. Always smoking dark red I cannot see anyone.
> Many people run. I just follow.

> Pretty soon like fog. Red fog. And gray. And people down all around me.
> And people look so awful. Skin comes off. Just awful. Makes me so scared, so afraid. I never knew such hurt on people. Un-human. I think, if I am in hell, it is like this.
> No faces. No eyes. Red and burned, all things, like women's hair, dusty and smoking with burning.
> Many people going to the river. I watch them. Many people drinking water. But they fall in and die, and they float away. Voices cry, calling names. I cannot hear because so many voices cry, all calling names. So many voices.[4]

I feel her voice to be "filling me with a precious essence"[5] that I cannot translate. What is she saying? It is not the broken English, spoken by a girl broken from the war that is a problem; it is what she is saying. It is incomprehensible. Unrepresentable. Unclassifiable (*atopos*). Unforgettable. Abstract, but moving: not unlike the abstract maps that make up this book.

Flowers

I remember the darkened classroom, the sound of the film moving through the sprockets of the reel, the constellation of bits of dust dancing in the projector's light (like a dandelion blown), the shock of seeing fellow students crying, even boys. Still, I can hear the words from the film, especially the bit about the atomic flowers: "And by the twentieth day after the bomb, vegetation began to grow wildly in the wreckage of the cities. From charred remnants of plants, lush green weeds and wildflowers sprung again, madly, in extraordinary regeneration, stimulated by atomic radiation. And as people died of radiation sickness, the cities were blanketed with flowers."[6]

John Hersey in his famed 1946 book, *Hiroshima*, also tells of the blossoming of flowers, the beauty, "the optimistic green" that sprang from the bomb:

> Over everything—up through the wreckage of the city, in gutters, along the riverbanks, tangled among tiles and tin roofing, climbing on charred tree trunks, was a blanket of fresh, vivid, lush, optimistic green; the verdancy rose from the foundations of ruined houses. Weeds already hid the ashes, and wild flowers were in bloom among the city's bones. The bomb had not only left the underground organs of plants intact; it had stimulated them. Everywhere were bluets and Spanish bayonets, goosefoot, morning glories and day lilies, the hairy-fruited bean, purslane and clotbur and sesame and panic grass and feverfew. Especially in a circle at the center, sickle-senna grew in extraordinary regeneration, not only standing among the charred remnants of the same plant, but pushing up in new places, among bricks and through cracks in the asphalt.

It actually seemed as if a load of sickle-senna seed had been dropped along with the bomb.[7]

Mine is a screen memory of a film that I saw when I was sixteen. I thought that I had forgotten all about it, until recently when the memory returned to me while looking at slavick's beautiful and terrifying watercolors of places that the United States has bombed. This memory came to me as an involuntary memory.

Proust coined this notion of "une mémoire involuntaire" (in his long *In Search of Lost Time*) against the more base experience of a "voluntary memory, the memory of the intellect."[8] (As Roland Barthes says of *his* jouissance, an echo of his beloved Proust's involuntary memory, "my body does not have the same ideas that I do."[9]) We have all experienced involuntary memories before: a sound, a taste, a smell, a touch, a sight that suddenly takes us on a trip back to the past. Involuntary memories, entirely driven by chance, find you; they cannot be planned. For Proust, it was, most famously, a nibble of madeleine cake soaked in lime blossom tea, bringing a rush of childhood memories back to him.[10] Memories that blossomed unexpectedly.

"Blossoming Bombs" is a story of the blossoming of an unexpected memory.

Memory

My Proustian memory happened when I was first looking at slavick's striking drawings that now make up the maps of this book: drowned in ink that has been plopped down on the paper, in an echo of the senseless repetition of dropping bombs. Scratched, smudged, layered like the residue of toppled buildings after an air strike, these maps are worthless for actual navigations. Without legends, without clear markings of any kind, they are, instead, maps for thinking or rethinking. In them, we get lost: which is, in this case, a good place to be. Slavick's dystopia is a long series of places that the United States has bombed: traces of horrors well-known and less-known, including *Hiroshima*, *Nagasaki*, *Laos*, *Iwo Jima*, *Normandy*, *Pakistan*, *Guatemala*, *Tokyo*, *Boken Island*, and *New Mexico*. There are sixty drawings. But there could be more. Depressing as that is, each drawing, whether or not it appears as green as *Pakistan* or *Proving Ground, Nevada I*, was made with "the optimistic green" of hope. The hope for world peace.

Yves Klein also used the sloshing of color on maps with the (blue) utopian hope of ending war. As Nan Rosenthal has written about Klein's solid blue relief map of 1961, *Europe-Afrique*: "Klein used color as though it could be an explicit and overtly political tool for ending wars, because if you paint a single color over a relief map of Western Europe and North Africa, you thereby eliminate the boundaries between the countries with a unifying bath of blue."[11] In tune with Klein's abstract cartography, slavick's maps also use color to end war, if from a less utopian plain than Klein's.

Emitted, for me, from slavick's colorful, worked over, muddy, and bright abstractions was a particular kind of sign, what Gilles Deleuze calls (in his particular reading of Proust) a "sensuous" sign: a sign that gives us "a strange joy at the same time that it transmits a kind of imperative."[12] Via this sensuous sign, I felt something intense, something bodily, but what was it? The intensity was there, but it was not yet developed: it would blossom into what I would later understand as a memory, dislodged, from my adolescent body. I had yet to "develop" this quality, this sensuous impression. Like a "tiny Japanese paper that opens under water and releases the captive form,"[13] the material for flowering was there, but would remain closed, if only for a little while. In Proust's exact words: "as in the game wherein the Japanese amuse themselves by filling a porcelain bowl with water and steeping in it little pieces of paper which until then are without character or form, but, the moment they become wet, stretch and twist and take on color and distinctive shape, become flowers . . . all the flowers in our garden."[14]

As a child, I as well had my own Japanese games. I used to buy a modern equivalent, akin to Proust's pretty metaphor, in the form of little discs of pressed paper that I found in the shops of San Francisco's Chinatown, shops that smelled appealingly foreign with incense-laden sweetness. Stacked full of cheap things for children to buy, giant painted wooden yo-yos, candy wrapped in paper that you could eat, paper fans, staplers in the shape of wooden animals, giant pencils with colorful tassels, and, of course, these wonderful paper flowers. As soon as I returned home, I would put the enchanting little paper Jack-in-the-beanstalk magical-beans into the bottom of a tall clear glass of water. In the morning, I would awaken to a glass turned tall pond of marvelous paper water flowers, red, blue, yellow, and deep pink, all grown from tender stems of green thread, blossoming from their captive form.

Proust's Japanese paper flowers and those of my childhood were exquisite and immediate, echoing the intense feeling of an involuntary memory, which is more pure, more sincere than the memories we intellectually strive for. They just grow out of nowhere, not unlike the satisfaction of making a sun print on blue cyanotype paper.

I see Anna Atkins's nineteenth-century cyanotype *Papaver* (FIG. 1).

Hiroshima

In slavick's *Hypocenter in Hiroshima*, polka dots of alabaster wool hover over a pink and gray map speaking a silent (but far from quiescent) sign language hailed by the city's blasted center. At the center there was no sound, but slavick has prettily and eerily marked the silence with the sound of color. This pattern of slavick's *Hypocenter* cartography could echo the decorative scheme of any woman's fashionable forties American dress, of those women who sat at home unknowingly day dreaming as *Little Boy* was dropped, as children (every bit as precious as their own) vanished in Hiroshima, or left their childish shadows, their little prints, on the stone steps of their school. The writer Marguerite Duras, in her screenplay *Hiroshima mon amour*, has referred to these shadow-images, like the famous nebulous silhouette of the unknown person who sat on the steps waiting for the Sumitomo Bank to open, only to vanish with the light of the bomb, as "deceitful pictures."[15] In the words of the

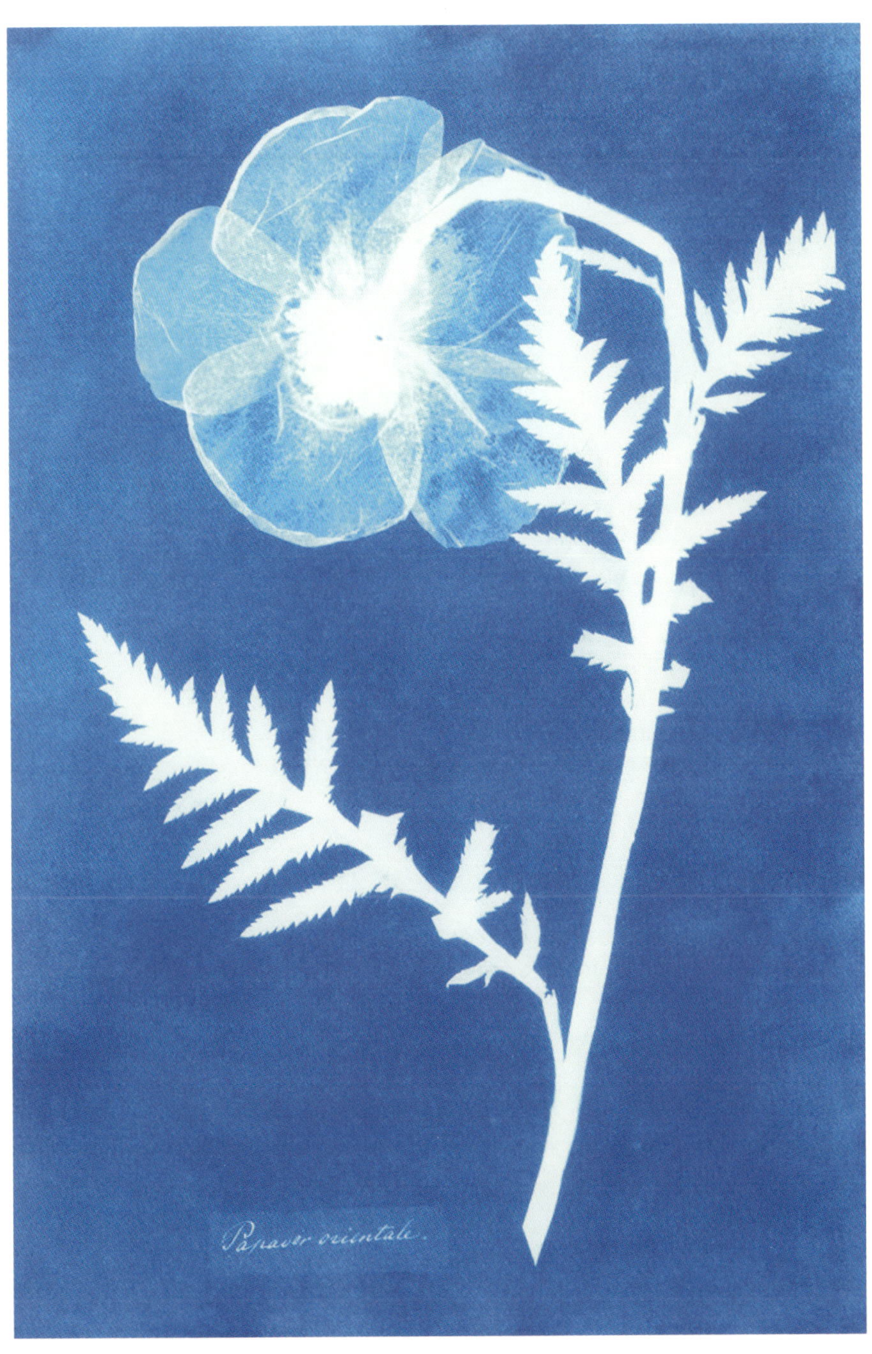

FIG. 1: Papaver (Oriental)
Anna Atkins, 1853, cyanotype
Victoria and Albert Museum, London

Hiroshima poet Tÿge Sankichi, this image of a vanished person, now forever in-waiting, becomes painted pattern: for me, his words hail the iconography of a slavick map:

> Enclosed by a painted fence
> on a corner of the bank steps,
> stained onto the grain of the dark red stone:
> a quiet pattern.[16]

And there are many such nuclear photographs: like the shadow left of the woman crossing a bridge over the river, wearing *monpe* or skirt pantaloons: the imprint of her vanished body captures the movement of her walking at the fatal second when the bomb dropped. She, like the bank patron, is nowhere to be found, not even buried in the ground, not even as fertilizer of the earth.

In 1953, Yves Klein saw the human shadow left on stone in front of the bank in Hiroshima. He was deeply moved and in 1961 made his own partially abstract image of barely visible shadows floating in his blue *Hiroshima*. As Klein wrote: "Hiroshima, the shadows of Hiroshima. In the desert of the atomic catastrophe, they were a witness, without doubt terrible, but nevertheless a witness, both for the hope of survival and for permanence—albeit immaterial—of the flesh."[17]

In *Hiroshima mon amour*, Duras also refers to the Hiroshima shadows as photographs on stone.[18] And, indeed, the Hiroshima photograms do conjure up photography's early days of shadow prints, not on stone, but on light sensitive paper, like Atkins's *Papaver*. (Fittingly, the poppy is a symbol of both memory and forgetting.) Curiously, Atkins's print looks like an X-ray: the first photograph to go into the body.

From within this blue poppy X-ray, I find myself back in the body of Proust. For, when Proust speaks of "involuntary words," he likens them to "a sort of X-ray photograph of the unimaginable reality which would be wholly concealed beneath a prepared speech."[19] Hiroshima's postnuclear photographs are images of an unimaginable reality. Vaporized by the bomb, we have nothing but these surreal shadows, sometimes white, sometimes black. The result is the unthinkable. Abstraction.

There can be no past and present (or future) in a body vanished. While Balzac may have feared losing thin ghosts of himself, like layers of skin, with each photograph "taken," with the Hiroshima photographs *everything is taken*.[20] These "deceitful" Hiroshima "photographs" are not indexical in a prenuclear sense, in which the referent is still felt to be present, as with the casts of victims caught in Pompeii's Mount Vesuvius eruption or, even, Atkins's photogram. The latter are still spatial: a cast taken, a flower put on paper. The totalizing loss of the referent in the Hiroshima "photographs" suggests the failure of indexicality in the postnuclear, a different moment of forgetting that could mean the whole world forgotten.[21] After the dropping of the A-bomb, all stories of bombing are necessarily soaked, splashed, scraped, scuffed, and scratched by the postnuclear. Slavick's maps, each and every one, hit us with the pan-abstraction of the possibility of total annihilation as radiated in the postnuclear age. Indeed, the focus of my essay could have been on any number of the tragedies that slavick maps. Unfortunately, *Hiroshima* is always alongside *Nagasaki*, which is alongside *Shifa*, which is alongside *Pakistan*, and on and on, bomb after bomb. There is no center.

A case in point is the Vietnam War, always shadowed by the postnuclear condition. Read this famous passage from Jean-Paul Sartre's *On Genocide* alongside slavick's *Laos*, *Vietnam*, *Hiroshima*, and *Nagasaki*:

> When a peasant falls in his rice paddy, mowed down by a machine gun, every one of us is hit. The Vietnamese fight for all men and the American forces against all. Neither figuratively nor abstractly. And not only because genocide would be a crime universally condemned by international law, but because little by little the whole human race is being subjected to this genocidal blackmail piled on top of atomic blackmail.[22]

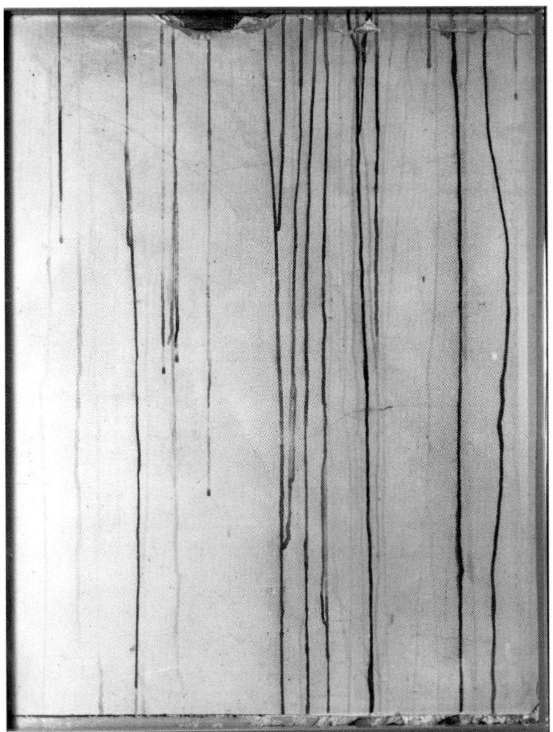

FIG. 2: Black Rain on White Wall,
Photograph by Hiromi Tsuchida, 1995
Hiroshima Peace Memorial Museum
Copyright Hiromi Tsuchida

Abstract

Jean Baudrillard has argued that the paintings of the postwar period, works by artists like Jackson Pollock, Robert Motherwell, and Clyfford Still, are in fact not cool, meaningless abstractions of purely formal Greenbergian concerns, but rather are hot meaningful reactions to the fate of representation: not only as the end of (figurative) representation in painting, but also as the end of representation itself, as threatened by the abstraction of the ever-looming, imageless, Cold War.[23] For Baudrillard, these paintings represent the impossibility of representing the annihilation of the world. Baudrillard sees such explosive gestures as representations of the unthinkable.

Likewise, slavick's albeit smaller, more tender, explosive marks map for us a similar defeat of representation, in hopes of nothing less, nothing more, than saving the world. With Baudrillard at hand, the compelling abstraction of slavick's *Kwajalein Atoll* finds itself alongside Hiromi Tsuchida's photograph, *Black Rain on White Wall* (FIG. 2). The chunk of wall is one of some 7,000 articles from the aftermath of the bomb that can be found in the Hiroshima Peace Memorial Museum.

I Do Not Understand

I do not understand. I cannot comprehend the fact that the famed silver monster plane dropped *Little Boy* on little boys and little girls, like the beautiful two-year-old Sadako Sasaki,[24] who would become the inspirational paper-crane-folding, speed running, kimono-wearing, peace-loving teenager who finally succumbed to her radiation-poisoning at

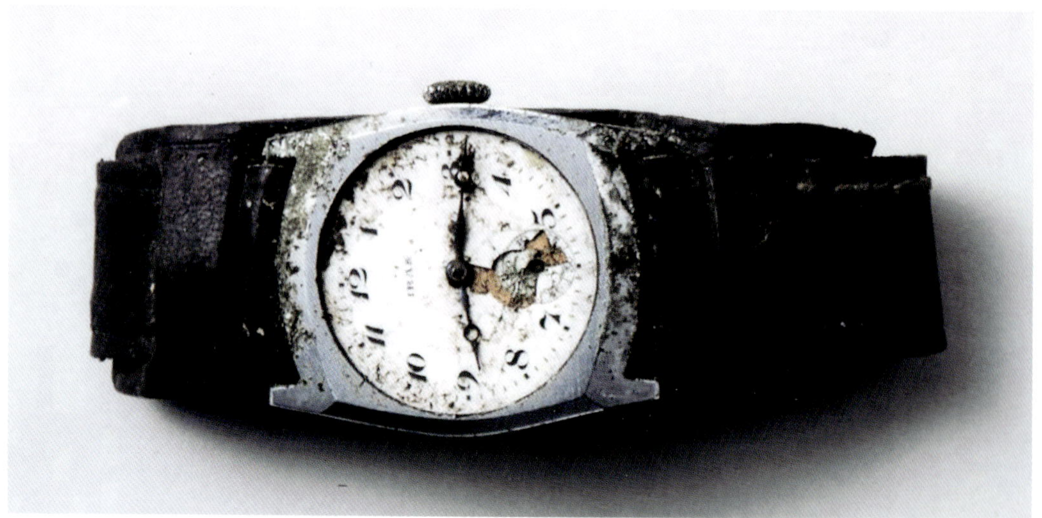

FIG. 3: Watch Stuck at 8:15
Photograph taken by the staff of the Hiroshima Peace Memorial Museum
Watch donated by Akito Kawagoe
Hiroshima Peace Memorial Museum

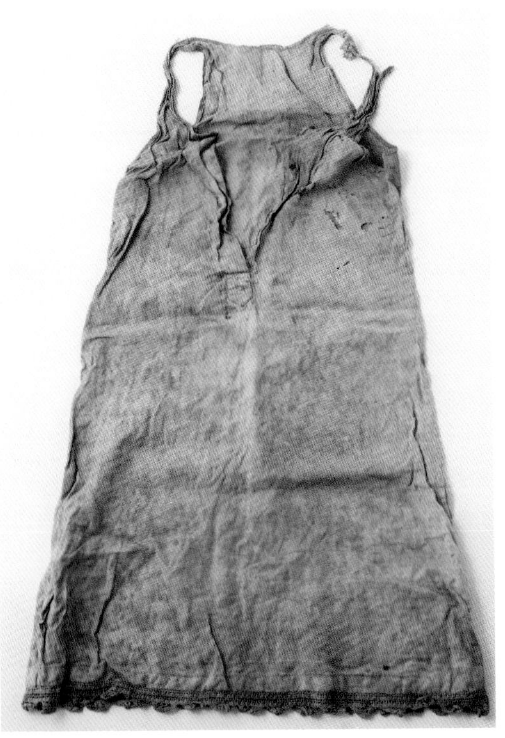

Fig. 4: Slip Stained with Black Rain
Photograph by Hiromi Tsuchida, 1995
Hiroshima Peace Memorial Museum
Copyright Hiromi Tsuchida

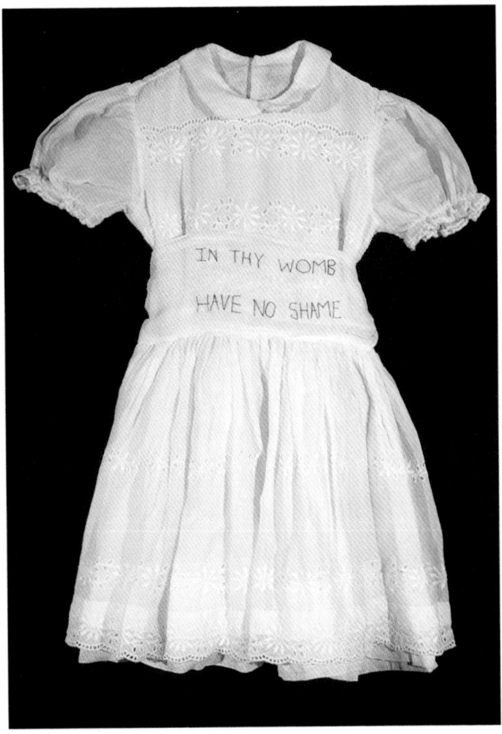

Fig. 5: In Thy Womb, Have No Shame, elin o'Hara slavick
Embroidered dress from slavick's childhood,
from *A Wall of Incoherent Dresses*, 1992
Artist's Collection

age fourteen. Obscene is the fact that the pilot of the plane, Paul Tibbets, had the name of his mother, Enola Gay, painted on the nose of the plane, just days before the mission. How could a plane named for one's mother kill so many mothers and children?

Amongst Tsuchida's photographs of rescued objects from victims of the blast, most of whom vanished, some of whom suffered terribly only to die later: there are eye glasses, lunchboxes, shirts, pants, dresses, watches forever stuck at 8:15 (Fig. 3). Unforgettable for me is the slip worn by a thirty-six-year-old woman who was nursing her child.

Aiko Sato was nursing her baby by the window of her home, 2,000 meters from the hypocenter. She was caught under a fallen chest of drawers together with her baby, and rescued by her two children, ten and twelve years old. Pieces of window glass sent flying by the blast tore holes in her slip. The black spots were made by the black rain, which fell while Mrs. Sato sought refuge.[25]

A memory. A memory, perhaps especially poignant to mothers. Slavick's mother went to Hiroshima, bought her the Tsuchida book, and inscribed it to her, in fast, moving, thick-felt-tipped cursive: "To Ellen, elin . . . Please keep me always in your Memory, memory!

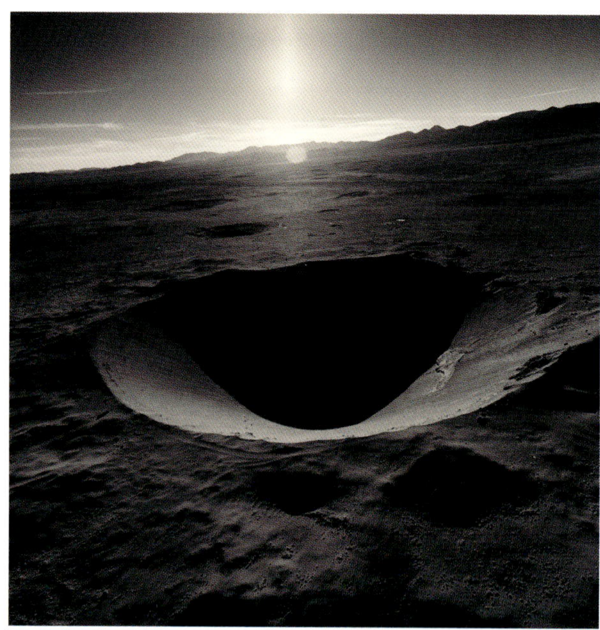

FIG. 6: Sedan Crater, Northern End of Yucca Flat, Nevada Test Site
Emmet Gowin, 1996
Copyright Emmet and Edith Gowin
Courtesy Pace / MacGill Gallery, New York

. . . The slip on page 102 is the most haunting image for me." (FIG. 4)

The Hiroshima slip recalls an early project of slavick's entitled *A Wall of Incoherent Dresses*, in which she sewed little poems on the dresses that she and her four sisters wore as children. Like a photograph, each dress holds traces of their childhood bodies, their precious past. The white communion dress, *In Thy Womb, Have No Shame*, is still mostly pristine (FIG. 5). No stains of black rain. Yet its message of the purity and the innocence of the womb exposes its vulnerability and that of the earth. Many unborn children in their mother's womb at the time of the blast would later face the shame of bodies twisted and scarred, of eyes that could not see, of minds that could not read and play.

Each page of *Bomb after Bomb* is an urging for us to look closely, page after page, at what we refuse to see: mothers, children, men waiting for banks to open. Likewise, slavick's book urges us to see what we cannot seem to see (to understand): that, to be silent is to be complicit in America's bombing of the world, a horrifying steady practice, in which even our own homeland is not forsaken.

As artist skydiver hovering above the earth, which we seem so intent on destroying, slavick holds hands with the photographer Emmet Gowin, also artist aeronaut. Both couple beauty with the unforgivable. Like the earth itself, slavick and Gowin give us beauty, so that we might just stop: stop to look; stop to think; stop bombing. Put Gowin's *Sedan*

Crater, Northern End of Yucca Flat, Nevada Test Site (Fig. 6), next to slavick's *Nevada Test Site II*, and listen to these words of Gowin's: "the astonishing thing to me is that in spite of all we have done, the earth still offers so much beauty, so much sustenance."[26]

Beauty

When I *first* looked at slavick's drawings, I was not reading titles. I was not getting all of this. I was just looking.

At first all I saw was the beauty: the layers of daffodil yellows, moss and forest greens, cranberry reds, taffy whites and tangerine oranges that animate the milky gouaches and rainy watercolors of these works. I felt the pleasures of summer. In some, like *We Are Our Own Enemy, Alamogordo, New Mexico* or *Baghdad* or *Mississippi*, I saw the cotton print dresses of my childhood. *Lebanon* was simply a beautiful butterfly. I felt sunshine in my belly. I held the marbles of my childhood, still warm with the summer of my life. I felt the prettiness of life. I remembered the first abstract painting that I had loved: Helen Frankenthaler's watery

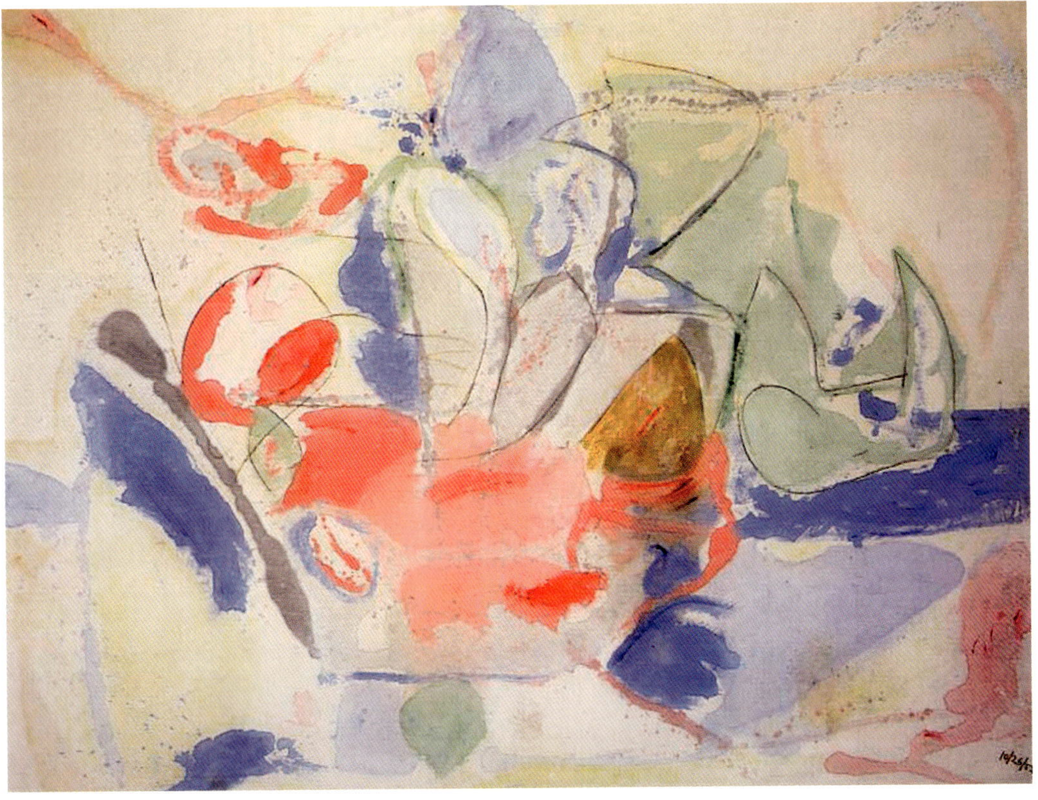

Fig. 7: *Mountains and Sea*
Helen Frankenthaler, 1952, oil on canvas, 7 x 10 feet

Mountains and Sea (Fig. 7). But when I took in the terrible words that accompany the maps of *Bomb after Bomb*, my joy was lost. Elation vanished as quickly as it had alighted.

Learning that the chrysanthemum petals of slavick's *New Mexico*, the red poppies of *Baghdad*, the purple bearded irises of *Mississippi* were gorgeous, symbolic marks illustrating not the bliss of the seasons, not the lightness of the pattern of summer dresses, not the warmth of the sunshine, not the purity of mountains and sea, not even the simple pleasures of abstraction, but rather places that the United States had bombed, my heart sank. I arched. I turned like milk, from sweetness to revulsion. I lost my color.

Like a dandelion burst, the sparkler was lit, the seeds released.

After seeing the cartography of violence in slavick's flowery abstractions of ghastliness, I managed to get my hands on a copy of the Iwasaki film. I watched its horrors on my home television, turning it on and off, as my eight-year-old son came in and out of the house. I did not want to scar him with the image of the Hiroshima woman whose skin had been dermagraphed by the design of her kimono from the heat flash of the bomb

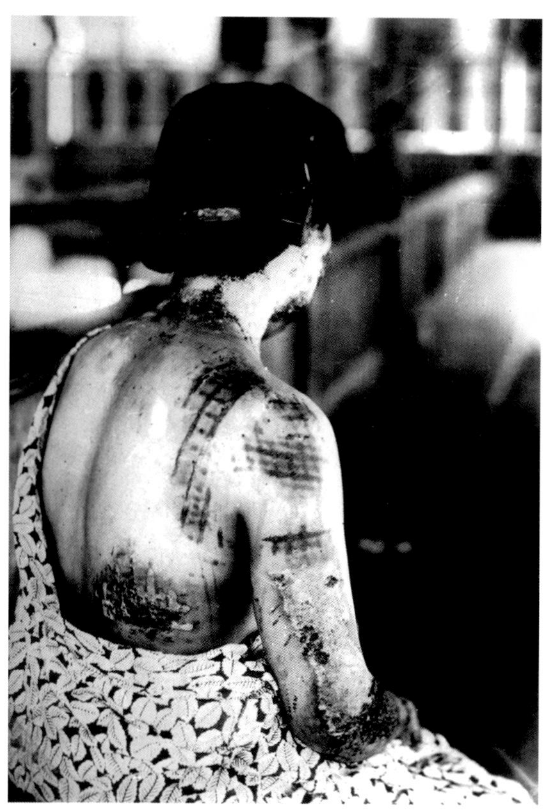

Fig. 8: Woman with Burns through Kimono, 1945
Photograph by Gon'ichi Kimura
Hiroshima Peace Memorial Museum

(FIG. 8). The dark areas of the pattern drew more heat and severely burned her skin. She was violently mapped with abstractions. Like the person waiting on the steps for the bank to open, this woman in a ravaged kimono with ravaged skin, she, too, was horribly photographed. Both victims are scarred by light, like the sixth-grader reading his own poem in Tÿge's "The Scar":

Little boy Okamoto finishes reading
Raises his short poem high with both hands
Finishes reading and suddenly bows his head
A head with a bald spot far too big
The destroyed house and his mother's death
Scars directly etched onto his five-year-old head
Now a sixth-grader
He has an "A-bomb bald spot" that others tease him for
. . .
And the future
Becomes that shiny bald scar
And gnaws into everyone's heart.[27]

Watching Iwasaki's footage, as an adult, gave me a host of peculiar feelings: repulsion and horror at the onset, but also nostalgia, even what Proust has described as "an exquisite pleasure . . . something isolated, detached . . . with no suggestion of its origin."[28] I attribute these happy feelings, so out of joint with what I saw before me, to thoughts of a journey from high school to adulthood, with all of the emotions of seeing the film *now* as a mother, *now* as a professor, *now* in a world that is, miraculously *still here*. Perhaps Fredric Jameson is right: "it is easier to imagine the end of the world than the end of capitalism."[29] While our earth is every bit as vulnerable as it was when I was sixteen (if not more so), it nevertheless has survived, has remained wholly round. As Gaston Bachelard affirms: "being cannot be otherwise than round."[30] I can watch the horrors of a documentary film on Hiroshima and still find this roundness.

I Am Looking for Summer inside a Black Marble

I remember.

I remember that hot summer of 1945. I remember my mother's beautiful "forties" floral dress, with its little yellow flowers with orange centers like small jelly drops, happy blooms stamped upon a champagne-colored background. I remember my mother's dress with its tight-fitting bodice and voluminous skirt that danced just below her sun-kissed knees, knees that echoed her sun-kissed shoulders. Freckles. Ankles of espadrilles.

I remember my mother's 1945 full red lips, her gold-button earrings, her incredible beauty, a young man's desire for her, her *L'air du temps* scent.

I see my mother's dresser with her *L'air du temps* perfume bottle, its famous bottle crowned with kissing doves, made of clear glass. Clear glass like the marbles that occupied my childhood. Coveted marbles that you could look through. They were not cat-eye marbles. They had nothing inside. Pure emptiness. We called them "purees."

Harold Edgerton's 1952, post-Hiroshima, rapatronic photograph of an atomic bomb test in the Nevada desert captured the detonation at the fire ball stage before it grew into a mushroom (FIG. 9). Edgerton's mind-blowing image appears as a huge marble in the dark. Like a puree from my childhood, the orb is clear with light-catching bubbles of air throughout. Its scale almost indeterminable, the ball could be a blown-up photograph of a marble or it could be a telescopic photograph of a distant planet. Its circular interiority takes me to Clarence White's 1903 *Drops of Rain* (FIG. 10). In the White photograph, the sphere holds springtime showers and the promise of primroses: an Edwardian garden ball, a gazing ball, a garden globe to ensure beauty and ward off evil. The young boy in the photograph had no idea that, within forty-two years, the world would never be the same. It would be showered with black rain.

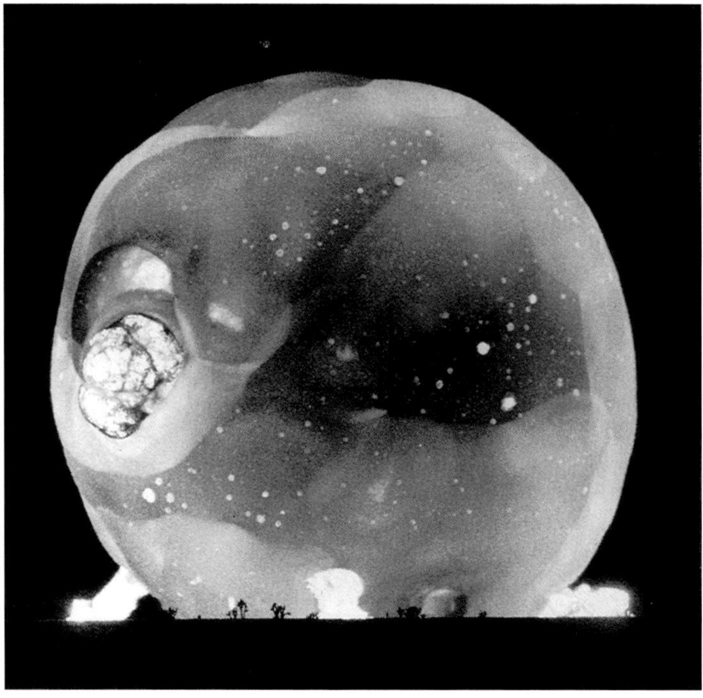

FIG. 9: Rapatronic Photograph of Nuclear Blast
Harold Edgerton, 1952
Copyright Harold and Esther Edgerton Foundation, 2007
Courtesy of Palm Press, Inc.

It is summer 1945.

I remember the bright sunny days cut by the fear of the war, the terrible fear that the world would end. At times, I could see it in my mother's anxious face. This "memory" is, of course, a "fantasy with a décor almost certainly derived from film"[31] and yellowed photographs. I was nothing more than a round egg waiting to enter her womb.

I saw nothing.

But, thanks to the buds and blooms of *Bomb after Bomb*, I remember. I remember, at least, to never forget.

Terrible Beauty

The terrible beauty that germinates in slavick's bomb drawings is where the flower petals curl the tightest: the heart. *Bomb after Bomb* is at home with some of the most poignant, yet poetic, artistic work on bombing. In the abstraction of slavick's paintings, there is an open-endedness in form and content. *Bomb after Bomb* is a book unbound by hope, an undoing that unlocks the grip of warfare and releases us as free pages. In slavick's heart as book, we are invited to contribute our own pages (leaves).

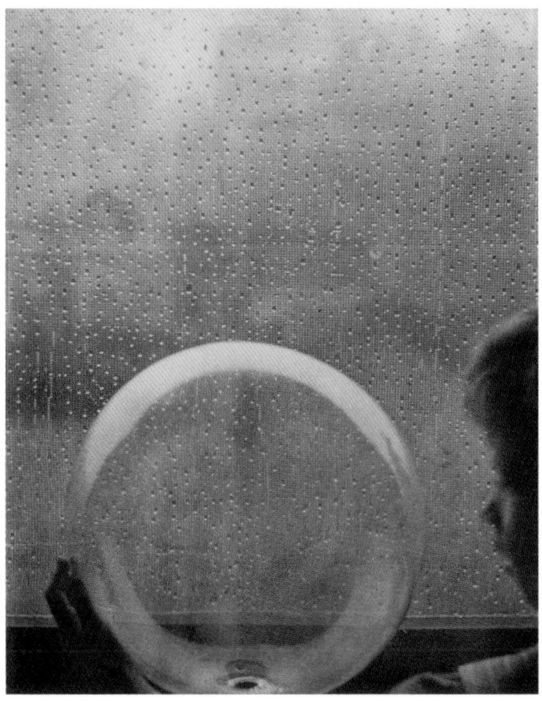

Fig. 10: Drops of Rain
Clarence White, platinum print, 1903
Copyright and Courtesy The Museum of Modern Art /
Licensed by SCALA / Art Resource, New York

While looking through a library copy of *Hiroshima and Nagasaki: The Physical, Medical, and Social Effects of the Atomic Bombings*,[32] I discovered something small, but extraordinary: I found four pressed pansies, intermittently spaced, between its pages (FIG. 11). Who left these scattered blossoms for me, for all readers of this book? Look at my favorite pansy amongst the four pages soothed, sweetened, and pacified with a flower. (How fitting that this purple, white, and dotted yellow pansy graces a page under the chapter heading of "Toward the Abolition of Nuclear Arms.") I love this flower that stares back at me, and now at you, with its pansy-face of hope. It, too, is a blossom "sprung again, madly, in extraordinary regeneration, stimulated by atomic radiation" (Barnouw, Iwasaki).

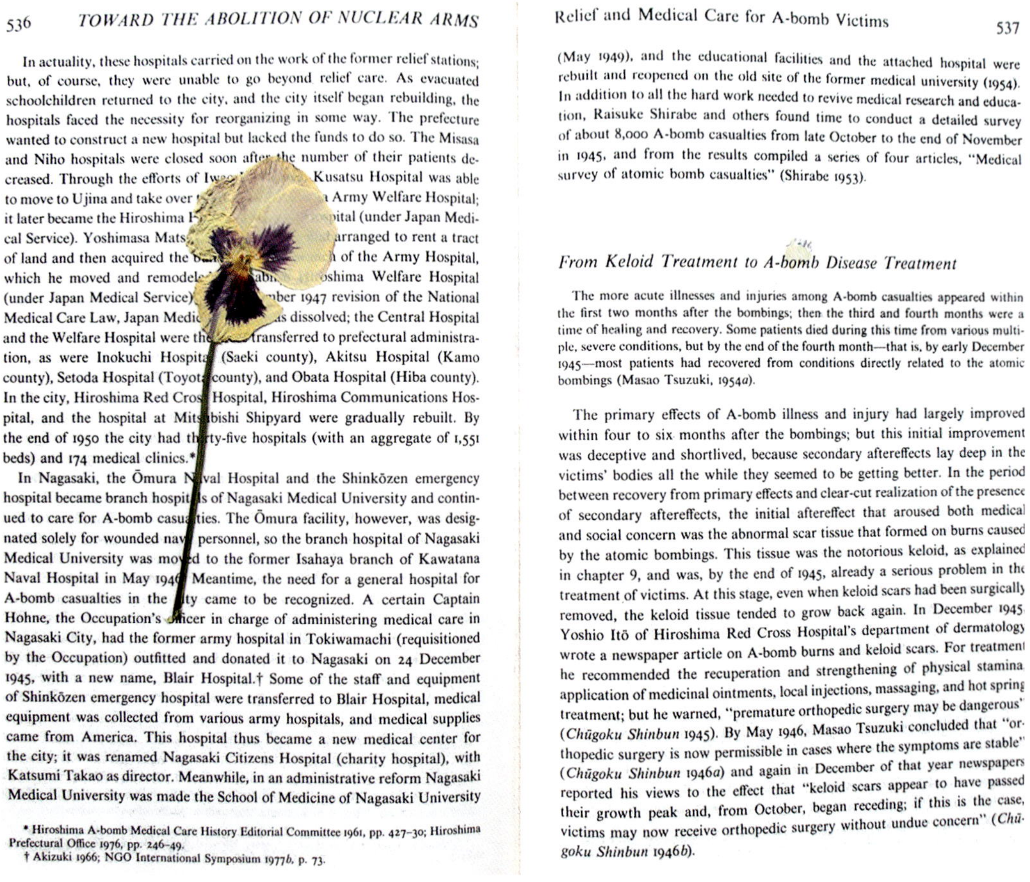

536 *TOWARD THE ABOLITION OF NUCLEAR ARMS*

In actuality, these hospitals carried on the work of the former relief stations; but, of course, they were unable to go beyond relief care. As evacuated schoolchildren returned to the city, and the city itself began rebuilding, the hospitals faced the necessity for reorganizing in some way. The prefecture wanted to construct a new hospital but lacked the funds to do so. The Misasa and Niho hospitals were closed soon aft[…]he number of their patients decreased. Through the efforts of Iw[…] Kusatsu Hospital was able to move to Ujina and take over […] Army Welfare Hospital; it later became the Hiroshima […]pital (under Japan Medical Service). Yoshimasa Mats[…] arranged to rent a tract of land and then acquired the […] of the Army Hospital, which he moved and remodel[…] oshima Welfare Hospital (under Japan Medical Service)[…]mber 1947 revision of the National Medical Care Law, Japan Medic[…]s dissolved; the Central Hospital and the Welfare Hospital were th[…] transferred to prefectural administration, as were Inokuchi Hospit[…] (Saeki county), Akitsu Hospital (Kamo county), Setoda Hospital (Toyot[…] county), and Obata Hospital (Hiba county). In the city, Hiroshima Red Cros[…] Hospital, Hiroshima Communications Hospital, and the hospital at Mits[…]bishi Shipyard were gradually rebuilt. By the end of 1950 the city had th[…]rty-five hospitals (with an aggregate of 1,551 beds) and 174 medical clinics.*

In Nagasaki, the Ōmura N[…]val Hospital and the Shinkōzen emergency hospital became branch hospit[…]s of Nagasaki Medical University and continued to care for A-bomb casu[…]ties. The Ōmura facility, however, was designated solely for wounded nav[…] personnel, so the branch hospital of Nagasaki Medical University was mo[…]d to the former Isahaya branch of Kawatana Naval Hospital in May 194[…] Meantime, the need for a general hospital for A-bomb casualties in the […]ty came to be recognized. A certain Captain Hohne, the Occupation's […]icer in charge of administering medical care in Nagasaki City, had the former army hospital in Tokiwamachi (requisitioned by the Occupation) outfitted and donated it to Nagasaki on 24 December 1945, with a new name, Blair Hospital.† Some of the staff and equipment of Shinkōzen emergency hospital were transferred to Blair Hospital, medical equipment was collected from various army hospitals, and medical supplies came from America. This hospital thus became a new medical center for the city; it was renamed Nagasaki Citizens Hospital (charity hospital), with Katsumi Takao as director. Meanwhile, in an administrative reform Nagasaki Medical University was made the School of Medicine of Nagasaki University

* Hiroshima A-bomb Medical Care History Editorial Committee 1961, pp. 427–30; Hiroshima Prefectural Office 1976, pp. 246–49.

† Akizuki 1966; NGO International Symposium 1977*b*, p. 73.

Relief and Medical Care for A-bomb Victims 537

(May 1949), and the educational facilities and the attached hospital were rebuilt and reopened on the old site of the former medical university (1954). In addition to all the hard work needed to revive medical research and education, Raisuke Shirabe and others found time to conduct a detailed survey of about 8,000 A-bomb casualties from late October to the end of November in 1945, and from the results compiled a series of four articles, "Medical survey of atomic bomb casualties" (Shirabe 1953).

From Keloid Treatment to A-bomb Disease Treatment

The more acute illnesses and injuries among A-bomb casualties appeared within the first two months after the bombings; then the third and fourth months were a time of healing and recovery. Some patients died during this time from various multiple, severe conditions, but by the end of the fourth month—that is, by early December 1945—most patients had recovered from conditions directly related to the atomic bombings (Masao Tsuzuki, 1954*a*).

The primary effects of A-bomb illness and injury had largely improved within four to six months after the bombings; but this initial improvement was deceptive and shortlived, because secondary aftereffects lay deep in the victims' bodies all the while they seemed to be getting better. In the period between recovery from primary effects and clear-cut realization of the presence of secondary aftereffects, the initial aftereffect that aroused both medical and social concern was the abnormal scar tissue that formed on burns caused by the atomic bombings. This tissue was the notorious keloid, as explained in chapter 9, and was, by the end of 1945, already a serious problem in the treatment of victims. At this stage, even when keloid scars had been surgically removed, the keloid tissue tended to grow back again. In December 1945 Yoshio Itō of Hiroshima Red Cross Hospital's department of dermatology wrote a newspaper article on A-bomb burns and keloid scars. For treatment he recommended the recuperation and strengthening of physical stamina, application of medicinal ointments, local injections, massaging, and hot spring treatment; but he warned, "premature orthopedic surgery may be dangerous" (*Chūgoku Shinbun* 1945). By May 1946, Masao Tsuzuki concluded that "orthopedic surgery is now permissible in cases where the symptoms are stable" (*Chūgoku Shinbun* 1946*a*) and again in December of that year newspapers reported his views to the effect that "keloid scars appear to have passed their growth peak and, from October, began receding; if this is the case, victims may now receive orthopedic surgery without undue concern" (*Chūgoku Shinbun* 1946*b*).

FIG. 11: Pressed Pansy Found in *Hiroshima and Nagasaki: The Physical, Medical, and Social Effects of the Atomic Bombings*
Photograph by elin o'Hara slavick, 2007
Artist's Collection

Terrible Beauty by Air

The coupling of beauty with war, especially aerial war, has always been part of the imagination of war. Long before the development of the atomic bomb, when bombing was still in its infancy, the brilliant character Saint-Loup, from Proust's *Search*, tells us of the ironic beauty of an air raid. Here, Proust is attempting to reveal the lethally seductive forces of air attacks to his readers. In the words of Saint-Loup:

> It is a magnificent moment when they [the attack planes] climb, when they fly off in *constellation*, in obedience to laws as precise as those that govern the constellations of the stars—for what seems to you a mere spectacle is the

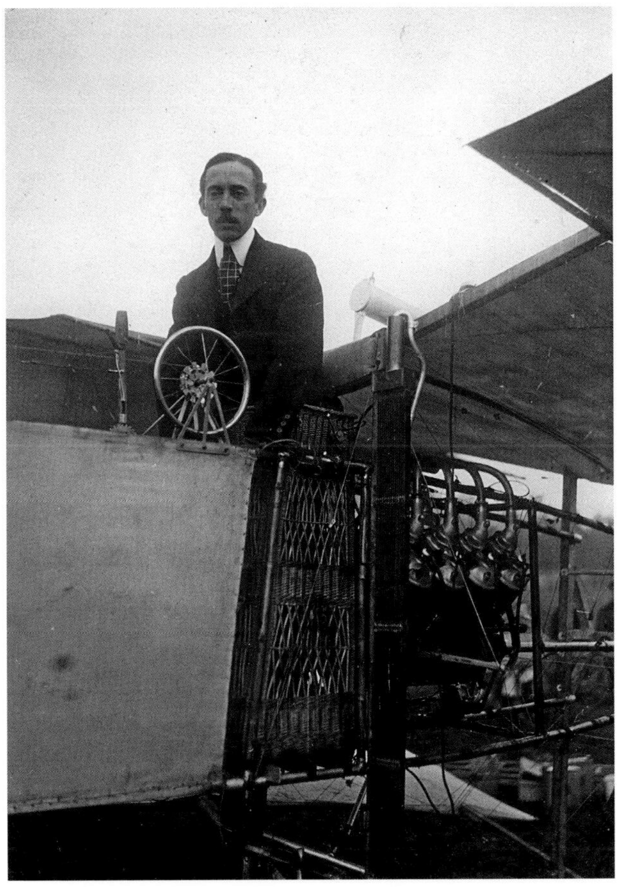

FIG. 12: Alberto Santos-Dumont in his N 14bis, France
Unknown Photographer, 1905
National Air and Space Museum Archives,
Smithsonian Institution, Washington D.C.

rallying of the squadrons, then the orders they receive, their departure in pursuit, etc. But don't you prefer the moment, when, just as you have got used to thinking of them as stars, they break away to pursue an enemy or to return to the ground after the all-clear, the moment of *apocalypse*, when even the stars are hurled from their courses? And then the sirens, could they have been more Wagnerian . . .?[33]

The dangerously awesome beauty of war is described everywhere, even with Hiroshima. Consider this passage from Diana's Preston's *Before the Fallout*:

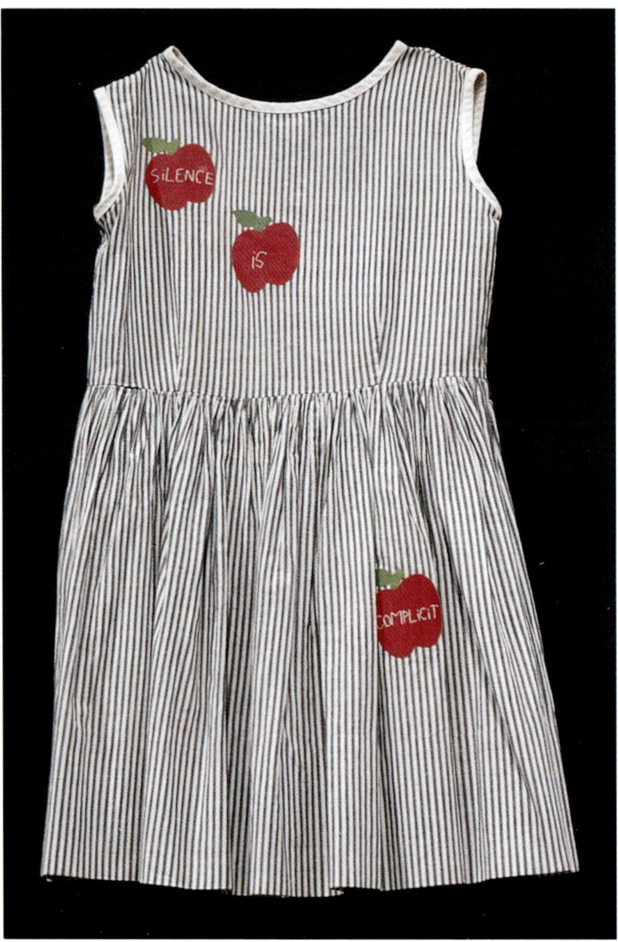

FIG. 13: Silence Is Complicit
elin o'Hara slavick
Embroidered dress from slavick's childhood,
from *A Wall of Incoherent Dresses* (1992)
Artist's Collection

On 6 August 1945, the Christian Feast of the Transfiguration, the Festival of Light, a young mother, Futaba Kitayama, looked up to see 'an airplane as pretty as a silver treasure flying from East to West in the cloudless pure blue sky.' Someone standing by her said, 'A parachute is falling.' Then the parachute exploded into an 'indescribable light.'[34]

Ironically, it is the airplane and its instruments that make all of slavick's maps possible. Before flight, only angels and birds could have such an awe-inspiring view. Airplanes have perfected cartography, marking the spots to be bombed; and airplanes are the carriers of bombs themselves. But flight was not always coupled with warfare.

For, to dream of flight before World War I (1914–1918) was a delicious thought far from death, couched in the height of the airy Belle Époque, a sweet soufflé before the time of its fall, before bombing sorties, before the British Royal Naval Air Service hit and destroyed the first Zeppelin on October 8, 1914.[35] To look at the tragically frail wooden frame of a Zeppelin shot down by WWI bomber planes is to look at an albatross of joyous flight turned dinosaur bones: "hurt and distraught . . . the flier's hobbled step" (Charles Baudelaire, "The Albatross").[36]

Flight's ironic beauty rings tragically true in the life of Alberto Santos-Dumont, whom much of the world, especially Brazil, considers to be the inventor of flight (Fig. 12). Santos-Dumont flew some two years before the Wright brothers.[37] (Brazil has long been outraged at America's nearly uncontested celebration of the Wright Brothers as the inventors of flight.)

On October 19, 1901, tiny 5'4" Santos-Dumont circled the spire of the Eiffel Tower in his innovative flying machine, making him the toast of Paris. Staying aloft for a full twenty-nine minutes and fifteen seconds, Santos-Dumont even managed to bring the airship with its magical rudders back to its starting point. When his workmen grabbed the guide rope and reeled him in like a yo-yo toy, Santos-Dumont "was showered with flower petals that swirled like confetti."[38] But the tiny aeronaut's flower-petal days would shrivel into deep depression with the onset of air combat, a despair that would last for his entire life. Eventually—on July 23, 1932, in a hotel room in Brazil—Santos-Dumont's grief would culminate into total darkness. After hearing a plane fly past and bomb a test target, the famed "Petite Santos" was found dead, hanging from "two bright-red ties from his flying days in Paris."[39] His last words were (spoken to the elevator operator): "I never thought that my invention would cause bloodshed between brothers. What have I done?"[40]

Slavick's drawings, inspired by the terrible beauty of flight and the terrible beauty of war, sow thoughts, words, colors, protests, feelings, memories like so many disseminating seeds, so as to get each and everyone of us, especially Americans, in this post-nuclear time to ask: "What have I done?"

After all, as slavick once stitched on the falling red appliquéd apples of a striped dress from her A *Wall of Incoherent Dresses*: "SILENCE IS COMPLICIT" (Fig. 13).

Bomb after Bomb is deafeningly loud and clear.

1. Barnouw was the first chief of Motion Picture, Broadcasting, and Recording Sound Division at the Library of Congress. The documentary was first shown in 1970 at the Museum of Modern Art. The struggle that Iwasaki (who had been taken to prison for his outspoken criticism of the Japanese government's new cinema law of the 1930s) had in shooting the film and keeping it safe from both the Japanese and American governments is a complicated affair. The complications included seeing his own footage, which he had assumed had been destroyed, in a broadcast of Barnouw's documentary on Tokyo television in 1970. Soon after that, Barnouw and Iwasaki would become friends and working partners for the remainder of their lives. See Erik Barnouw, *Media Marathon*. Durham–London: Duke University Press, 1996, pp. 193–217. Film enthusiasts who know the difficulty of saving a print of the footage from both governments will note the surprising appearance of the Iwasaki footage in the incredible film by Alain Resnais and Marguerite Duras, *Hiroshima mon amour* (1959).

2. Barnouw, Ibid., pp. 193–194.

3. The words of the Hiroshima girl are read by Kazuko Oshima, who had been in Japan during the war. At the time, she was a graduate student in the department of Film, Radio, and Television at Columbia University, for which Barnouw was Chair. Paul Ronder, who was also a graduate student at that time, is the voice of the male narrator. Ronder, along with Geoffrey Bartz, also did the editing. Barnouw, Ibid., pp. 194–195.

4. From Barnouw and Iwasaki's *Hiroshima-Nagasaki, August 1945*. I would like to thank Mara West for transcribing the film for me.

5. Marcel Proust, *Swann's Way*, of *In Search of Lost Time*, trans. C.K. Scott Moncrieff and Terence Kilmartin, revised by D.J. Enright. New York: Random House, 1992, vol. I, p. 60. There are six volumes to this edition.

6. As transcribed from the film, *Hiroshima-Nagasaki, August 1945*.

7. John Hersey, *Hiroshima*. London: Vintage, 1989, pp. 69–70.

8. Proust, *In Search of Lost Time*, *Op. cit.*, I, p. 59.

9. Roland Barthes, *Pleasure of the Text*, trans. Richard Miller. New York: Hill and Wang, 1975, p. 17.

10. There were also other things that sent Proust's Narrator travelling back through time: a flash of his grandmother a year after her burial while buttoning his boots; Venice while tripping on uneven paving stones; his uncle's sitting room through the cool musty smell of a latrine in the Champs-Élysées; a train ride through the chance knock of a spoon against a plate.

11. Nan Rosenthal, "Assisted Levitation: The Art of Yves Klein," in *Yves Klein, 1928–1962, A Retrospective*, catalogue from the exhibition at Rice Museum, Houston, February 5–May 2 1982. New York: The Arts Publisher, 1982, p. 123.

12. Gilles Deleuze, *Proust and Signs: The Complete Text*, trans. Richard Howard. Minneapolis: University of Minnesota Press, 2004, p. 11.

13. Deleuze on Proust's famous passage in which all of Combray (his boyhood home) comes rising from his teacup, Ibid., p. 11.

14. Proust, *Op. cit.*, I, p. 64.

15. Marguerite Duras, *Hiroshima mon amour*, trans. Richard Seaver. New York: Grover Press, 1961, p. 8.

16. Tÿge Sankichi, "The Shadow," as published in John Whittier Treat, *Writing Ground Zero: Japanese Literature and the Atomic Bomb*. Chicago: University of Chicago Press, 1995, p. 186.

17. Yves Klein, in Sidra Stich, *Yves Klein*. Stuttgart: Edition Cantz, 1995, p. 179.

18. Duras, *Hiroshima mon amour*, *Op. cit.*, p. 23.

19. Proust, *Within a Budding Grove*, *In Search of Lost Time*, *Op. cit.*, II, p. 222. In French, the passage is: "Ces paroles, de la sorte qui est la seule importante, involontaires, nous donnant la radiographie au moins sommaire de la réalité insoupçonnable que cacherait un discours étudié." See Proust, *À l'ombre des jeunes filles en fleurs*, in *À la recherché du temps perdu*, ed. Jean-Yves Tadié. Paris: Éditions Gallimard, 1988, vol. I, p. 577. So here in the original French, the X is not there, but the notion of the X-ray, the "radiographie," as the most piercing, most probing, form of photography is retained.

20. Nadar addresses this in his discussion of Balzac and the daguerreotype: "According to Balzac's theory, all physical bodies are made up entirely of layers of ghostlike images, an infinite number of leaflike skins laid one on top of the other. Since Balzac believed man was incapable of making something material from an apparition, from something impalpable—that is, creating something from nothing—he concluded that every time someone had his photograph taken, one of the spectral layers was removed from the body and transferred to the photographs. Repeated exposures entailed the unavoidable loss of subsequent ghostly layers, that is, the very essence of life." Nadar, "My Life as a Photographer," trans. Thomas Repensek, *October* 5, summer 1978, p. 9.

21. Ranji Khanna has been instrumental in helping me to understand the idea of the postnuclear and its relationship to representation. Perhaps most importantly, Ranji has shown me the value of a postnuclear reading of Proust and the ways in which Proust can open up a more politicized understanding of abstract painting after 1945. Hovering alongside hers is the work of Srinivas Aravamudan and his important writing on Oppenheimer's aestheticization of the blast of the bomb, describing its effect with a Hindu sacred text: "the radiance of a thousand suns were burst into the sky." I heard a lecture version of Aravamudan's "The Hindu Sublime, or Nuclearism Rendered Cultural," at the University of North Carolina, Chapel Hill, many

years ago and it had a profound effect on my understanding of the dropping of the bomb. See his *Guru English: South Asian Religion in a Cosmopolitan Language*. Princeton: Princeton University Press, 2006, pp. 142–183.
22. Jean-Paul Sartre, *On Genocide and a Summary of the Evidence and the Judgments of the International War Crimes Tribunal by Arlette El Kaïm-Sartre*. Boston: Beacon Press, 1968, p. 84.
23. Jean Baudrillard, "Hot Painting: The Inevitable Fate of the Image," trans. Richard Miller, in Serge Guilbaut, *Reconstructing Modernism: Art in New York, Paris, and Montreal 1945–1964*, ed. Serge Guilbaut. Cambridge, Massachusetts–London: MIT Press, 1992), pp. 17–29.
24. Sadako was but two years old when the bomb went off. The blast blew her out of the house, but she escaped without a burn or injury. Soon after the explosion, flames began to leap up in the area, causing Sadako's mother to flee carrying her two-year-old daughter in her arms. Near Misasa Bridge, they were caught by the black rain that came in drops the size of marbles. We can never really know what they saw, what they heard, what they smelled.
For a long time, Sadako was healthy and was famous for being a very fast runner. The radiation had chosen to sleep in Sadako, not waking up until she was ten years old, when she was diagnosed with "atomic disease" (leukemia). While in the hospital, Sadako made it a goal to fold 1,000 paper cranes, in order to verify the legend that by doing so a sick person will become healthy. At the top of Hiroshima's Children's Peace Monument dedicated to her, Sadako is ready to take flight.
Sadako worked hard at her cranes and brought joy and hope to everyone around her. But as she became sicker and sicker the cranes became tinier and tinier, as if mirroring her heart and her hope. So little, she had to begin shaping them with a needle, in order to make the tiny folds. So miniscule, the itty bitty cranes were no bigger than a few grains of rice. Although she surpassed her goal of making 1,000 cranes, she died on October 25, 1955, after only 4,675 days of life. So small, so much emerged from Sadako. Sadako was buried in a coffin filled with flowers, with only her face and hands peering out. She was blanketed in flowers, like Hiroshima after the bomb.
When Sadako's mother learned of her daughter's grave illness, she bought silk fabric with cherries on it and made her a beautiful kimono.
25. Hiromi Tsuchida, *Hirsohima Collection*. Hiroshima: Hiroshima Memorial Peace Museum, 1995, p. 102.
26. Emmet Gowin, in Jock Reynolds, *Emmet Gowin: Changing the Earth*. New Haven: Yale University Art Gallery, 2002, p. 155.
27. Tÿge, "The Scar," in John Whittier Treat, *Writing Ground Zero, Op. cit.*, pp. 184–185.
28. Proust, *Swann's Way, In Search of Lost Time, Op. cit.*, I, p. 60.
29. Fredric Jameson, *Archaeologies of the Future*. New York: Verso, 2005, p. 199.
30. Gaston Bachelard, *Poetics of Space*. New York: Beacon Press, 1994, p. 234.
31. Victor Burgin, *The Remembered Film*. London: Reaktion Books, 2004, p. 15. In this book, Burgin demonstrates how our memory is made up of a potpourri of real events and films that we have seen, the latter often becoming indistinguishable from "reality."
32. *Hiroshima and Nagasaki: The Physical, Medical, and Social Effects of the Atomic Bombings*. New York: Basic Books, 1981.
33. Proust, *In Search of Lost Time, Op. cit.*, IV, p. 54.
34. Ibid., p. 1.
35. Chris Chant, *The Illustrated History of the Air Forces of World War I and World War II*. London–New York–Sydney–Toronto: Hamlyn, 1979, p. 81.
36. Charles Baudelaire, "The Albatross," in *The Flowers of Evil* in French and English, translated by James McGowan with an introduction by Jonathan Culler. New York: Oxford University Press, 1998, p. 15. Originally published in French as "L'albatros" in *Les Fleurs du Mal* (1857). In French the phrase is, "maladroits et honteux…l'infirme qui volait;" see "L'albatros" in *The Flowers of Evil*, p. 14. Albatros was the name of the German army's most productive aircraft firm in 1914. See John H. Morrow, Jr. *The Great War in the Air: Military Aviation from 1909 to 1921*. Washington D.C.–London: Smithsonian Institution Press, 1993, p. 37.
37. On December 17, 1903, in Kitty Hawk, North Carolina, Wilbur and Orville Wright launched their Wright Flyer: an erratic 605 pound pterosaur, sporting a wingspan of forty feet, four inches, and a cheeky four-cylinder, water-cooled engine.
38. Paul Hoffman, *Wings of Madness: Alberto Santos-Dumont and the Invention of Flight*. New York: Theia, 2003, p. 140.
39. Ibid., p. 311.
40. As cited by Hoffman in Ibid., p. 310.

History is amoral: events occurred.
But memory is moral;
what we consciously remember is what our conscience remembers.
If one no longer has land, but has memory of land,
then one can make a map.

Anne Michaels, *Fugitive Pieces*

World Map, Protesting Cartography:
Places the United States Has Bombed, 1854–Ongoing

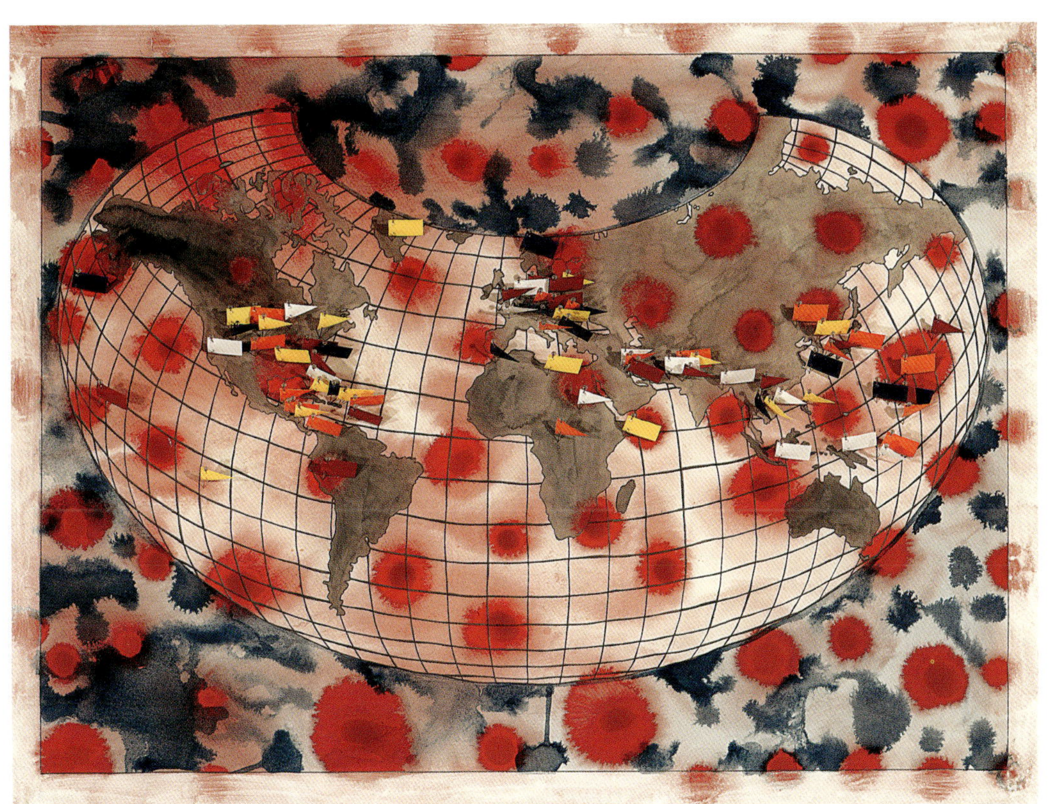

Nicaragua, 1854, 1928, and 1981–1990

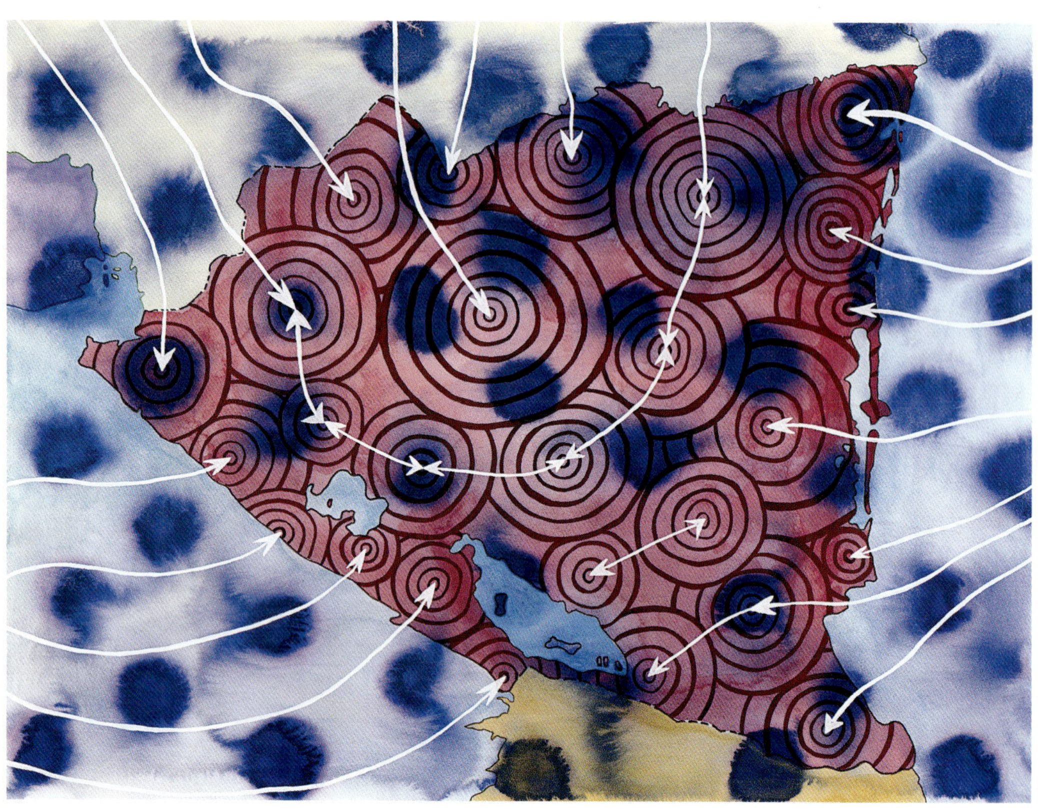

Firebombing of Dresden, Germany, 1939–1945

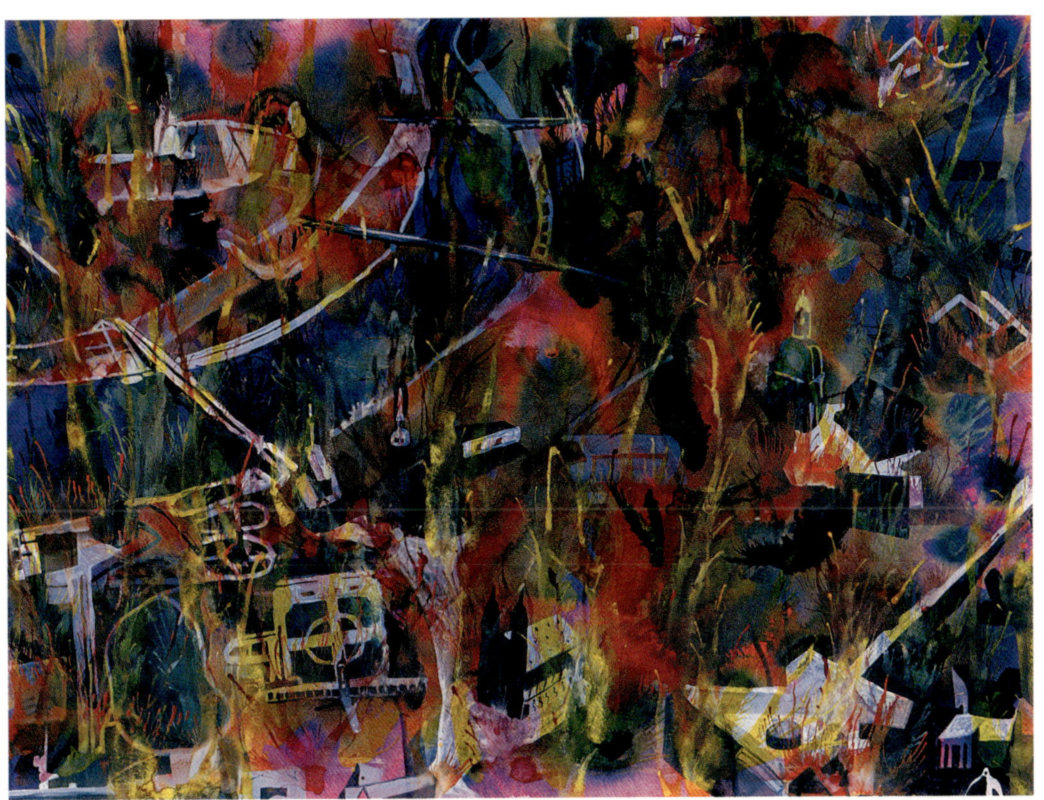

Vieques Island, Puerto Rico, U.S., 1941–1999

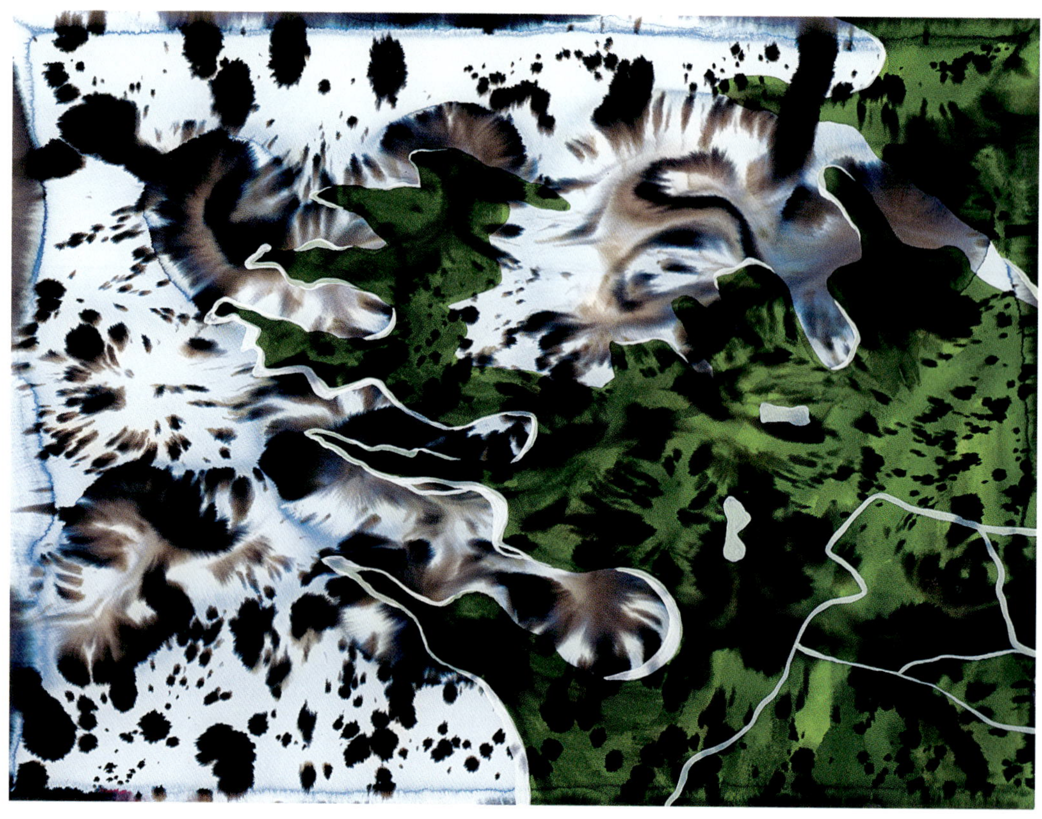

Dugway Proving Ground, Utah, U.S., 1942–Ongoing

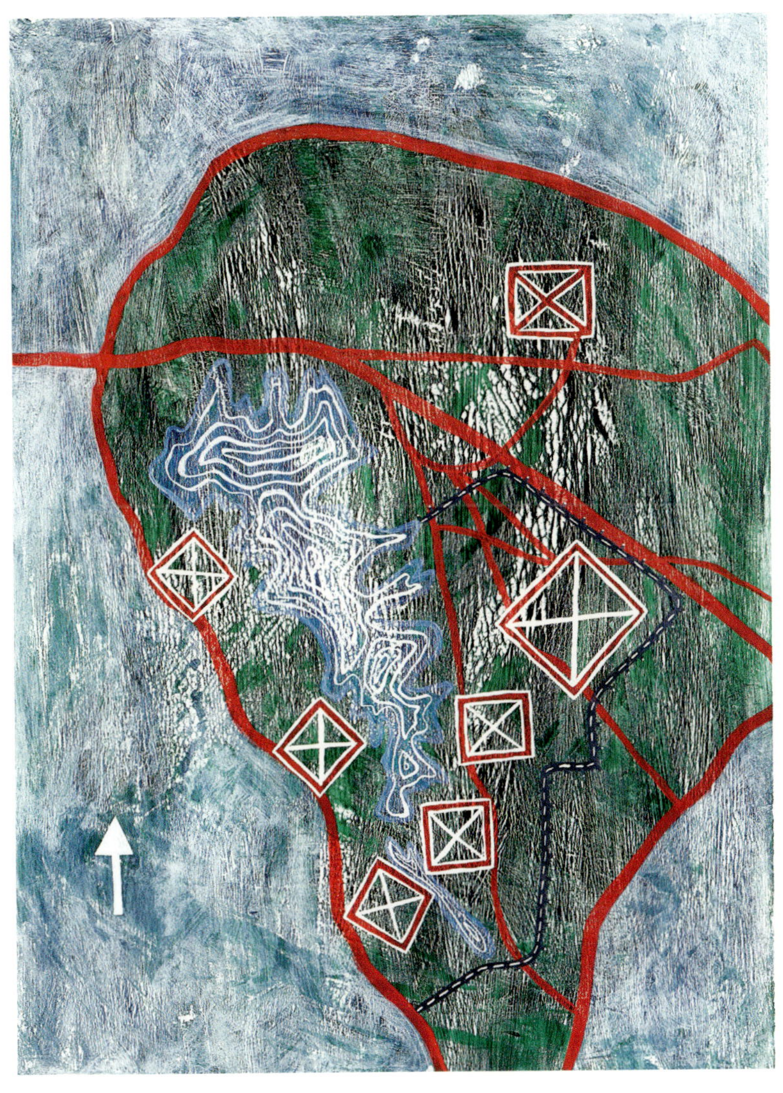

Solomon Islands, Operation Cartwheel, 1942–1945

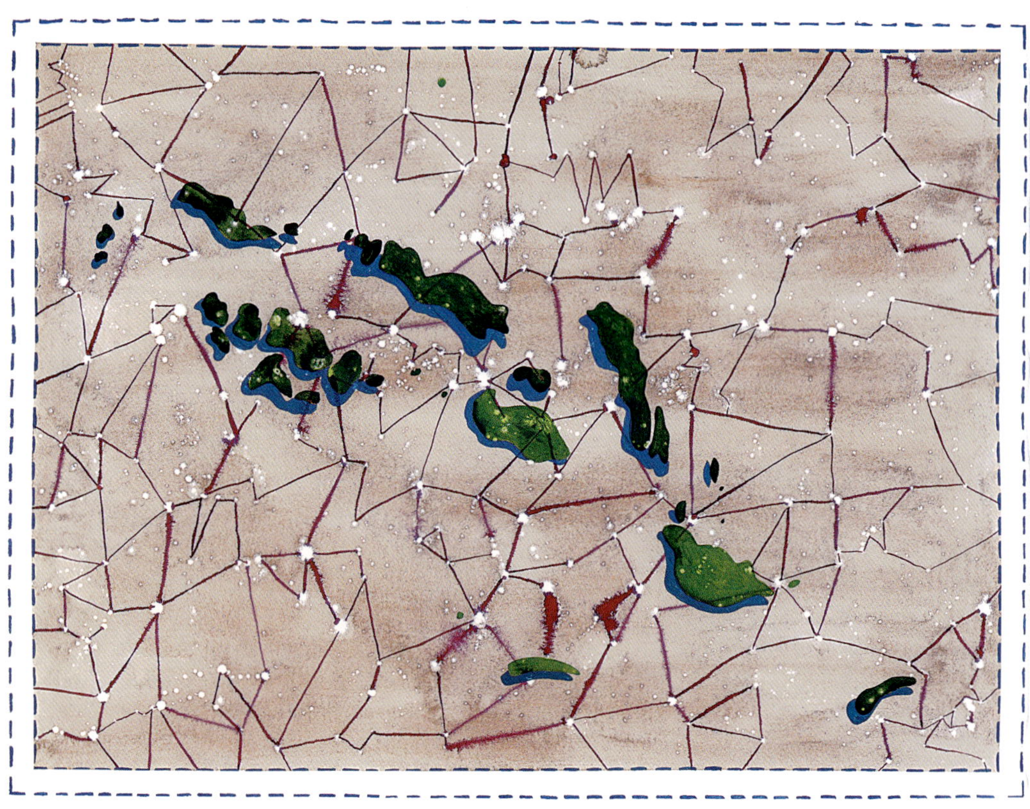

Nauru Island in the Gilberts, now the Republic of Nauru, 1943

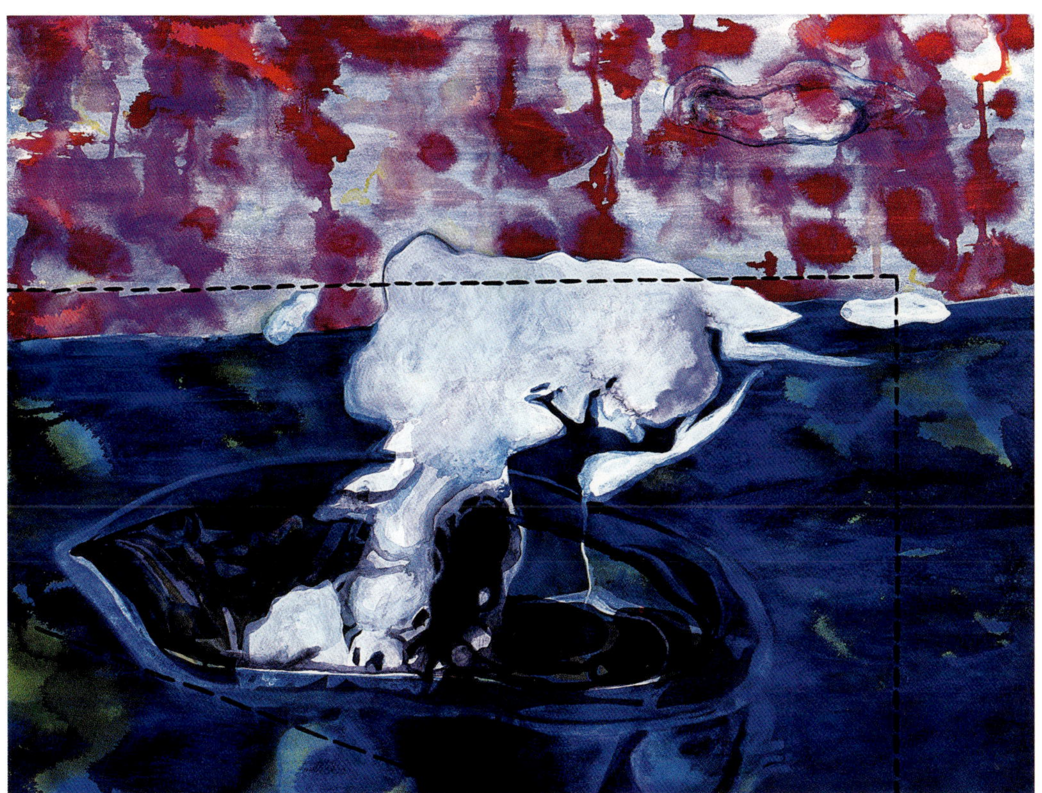

Poland, 1943–1944

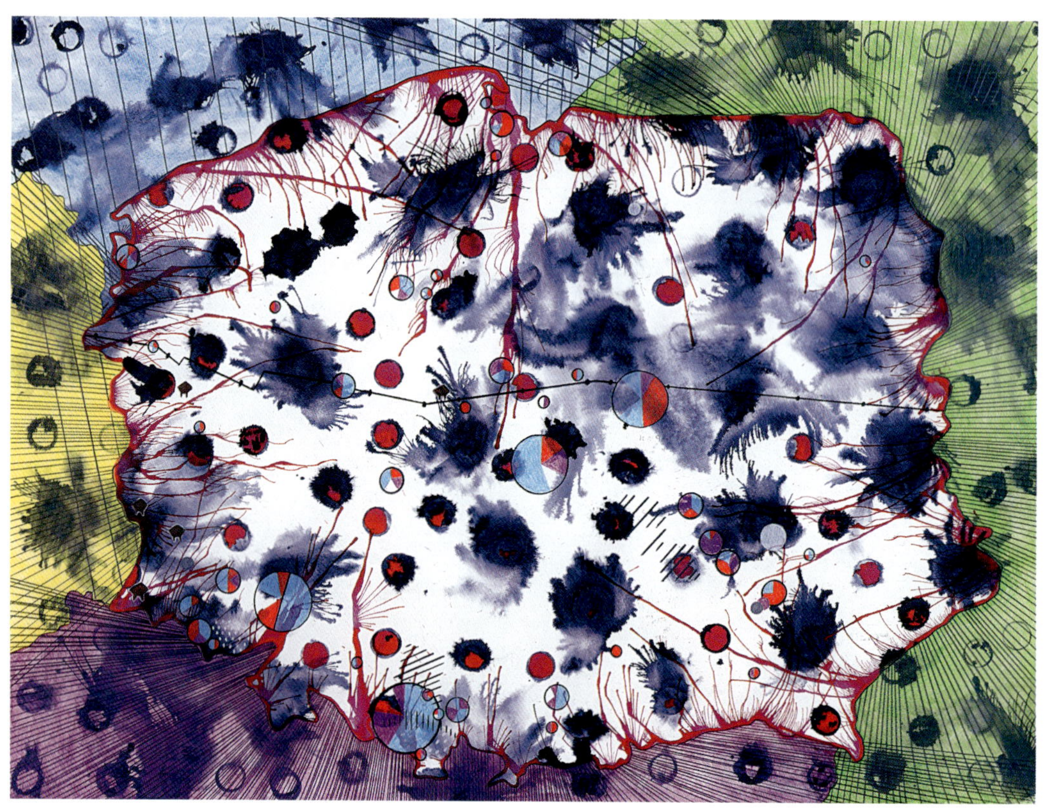

Assault on Iwo Jima, Japan, 1944–1945

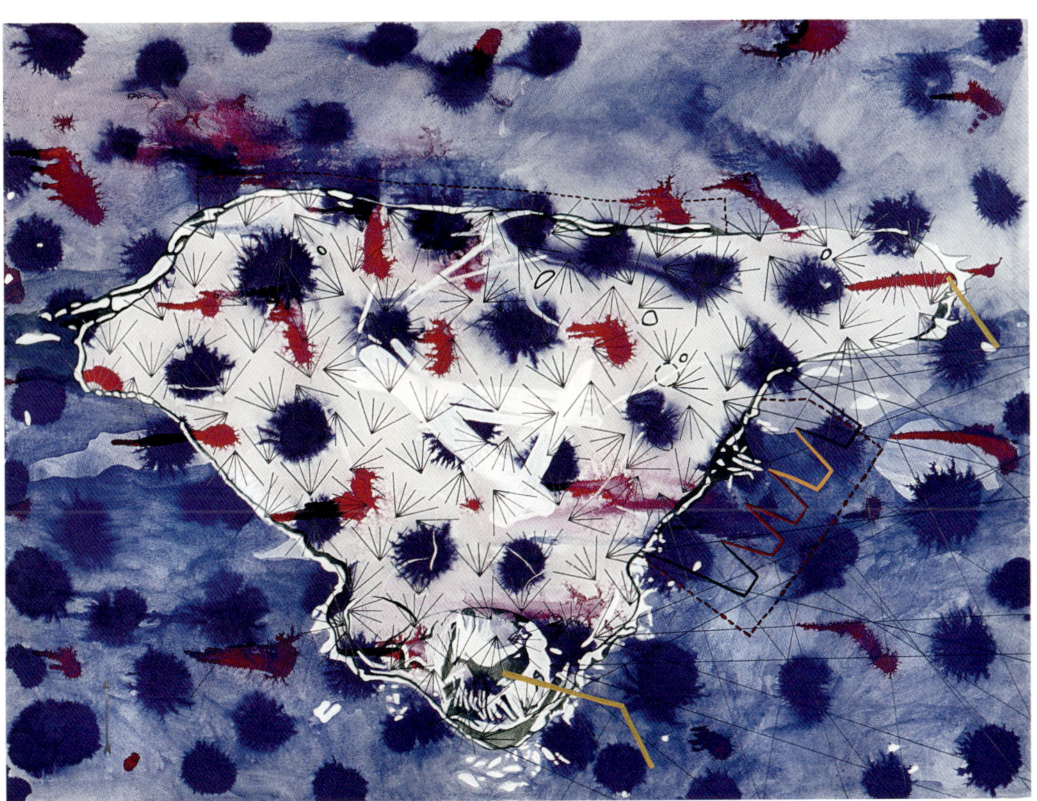

Kwajalein Atoll, Republic of the Marshall Islands, 1944–Ongoing

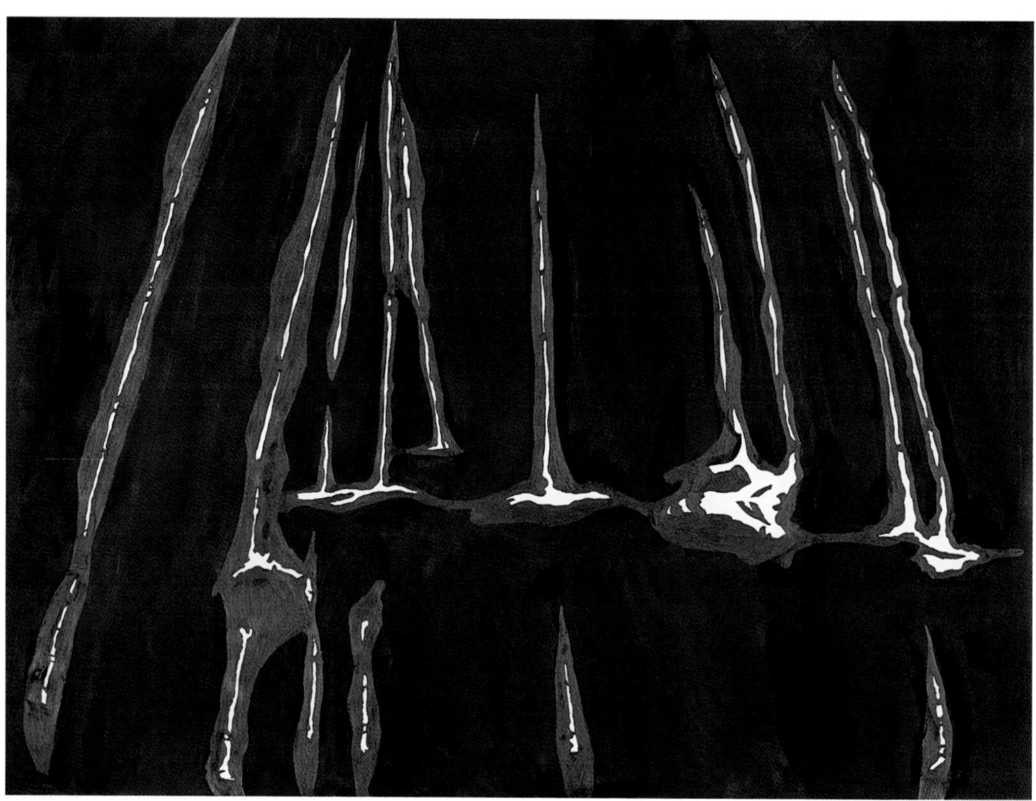

D-Day or Invasion Beaches, Normandy, France,
Operation Overlord, 1944

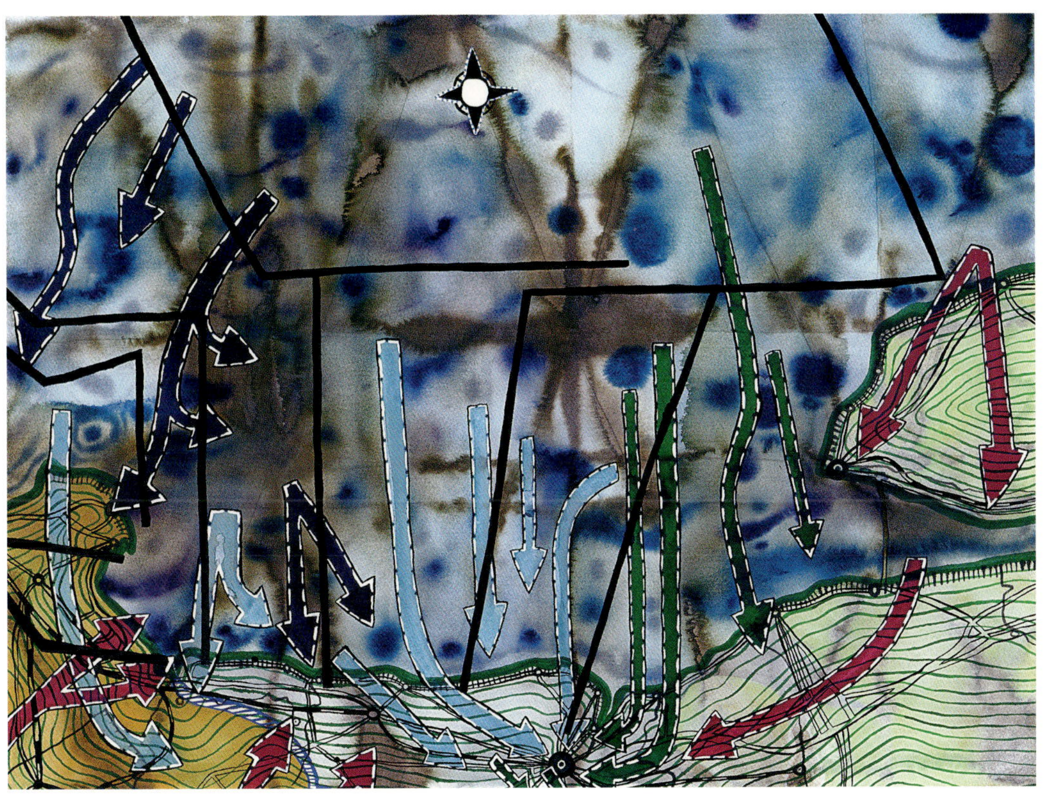

France, 1944–1945

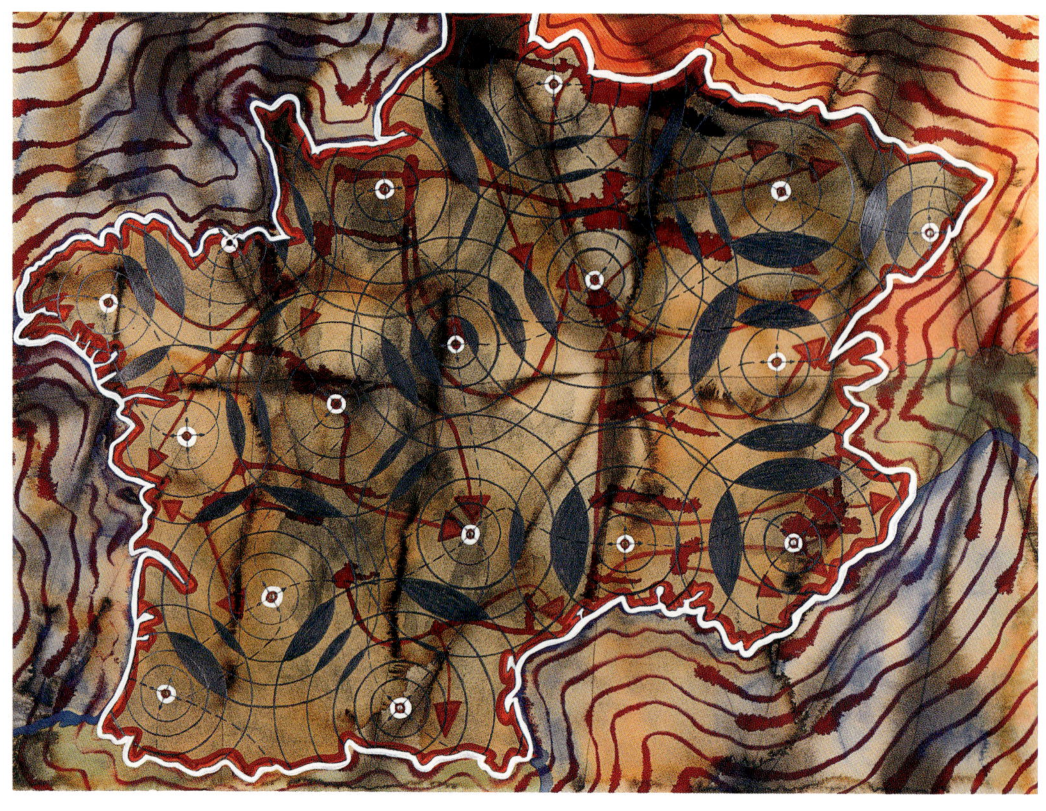

We Are Our Own Enemy, Alamogordo, New Mexico, U.S., 1945

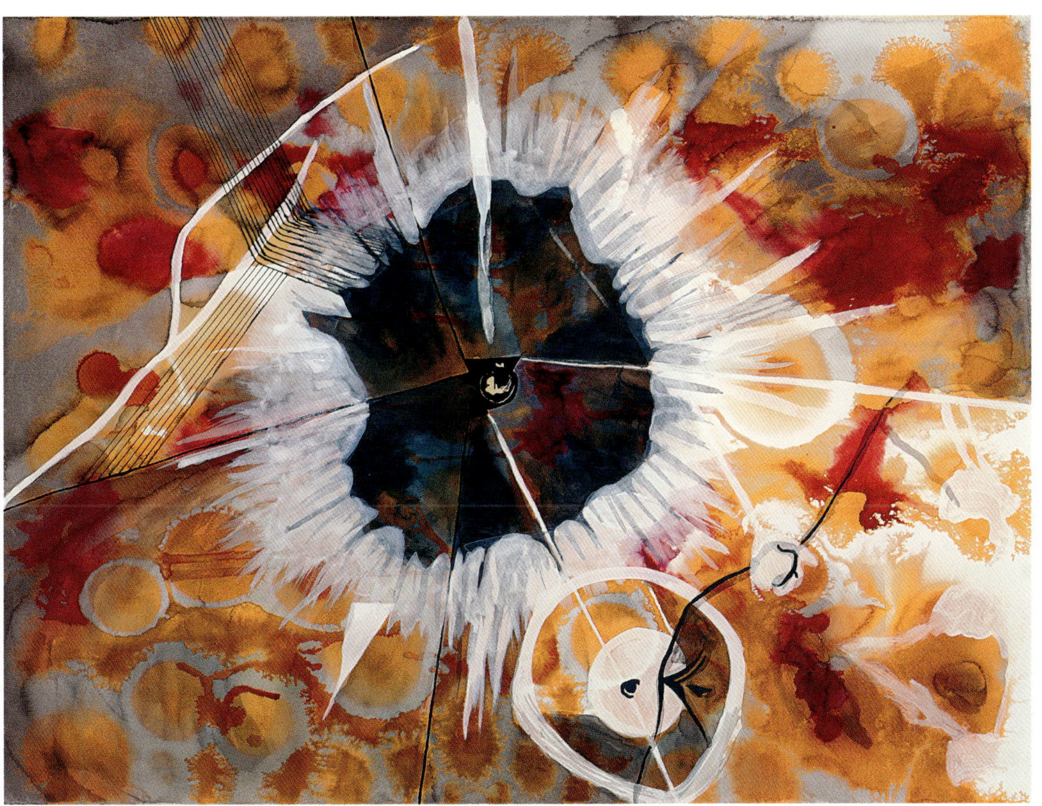

Hypocenter in Hiroshima, Japan, 1945

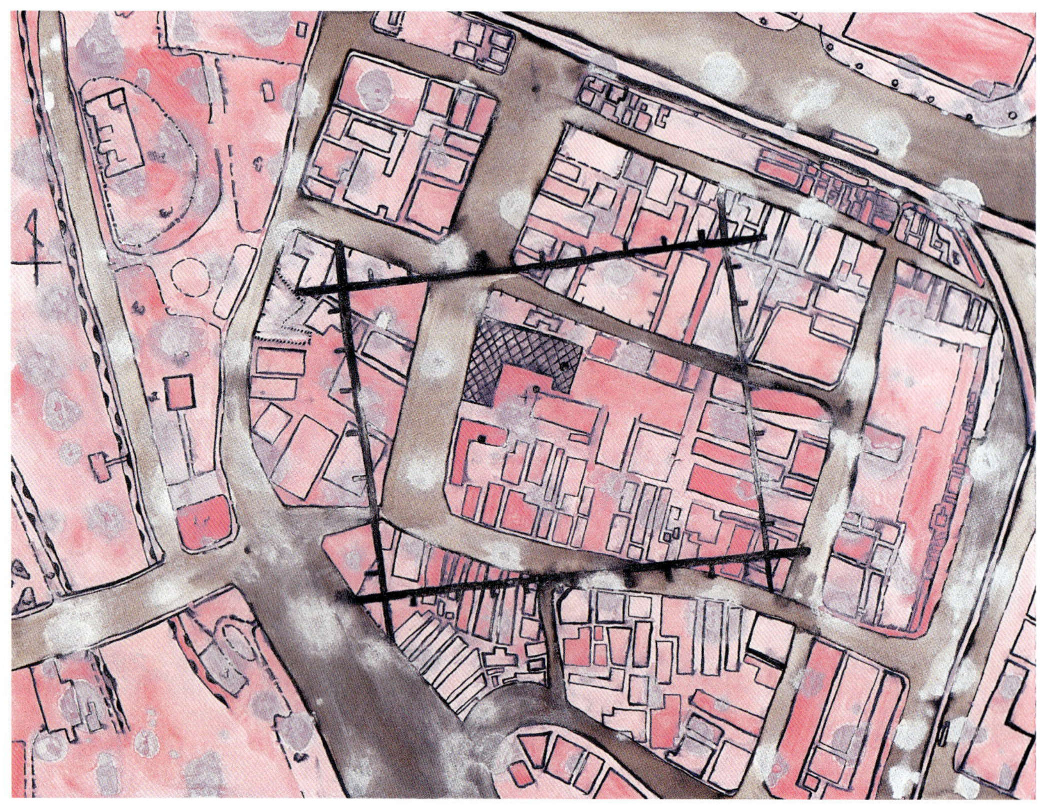

Nagasaki City at Bombing Time, Japan, 1945

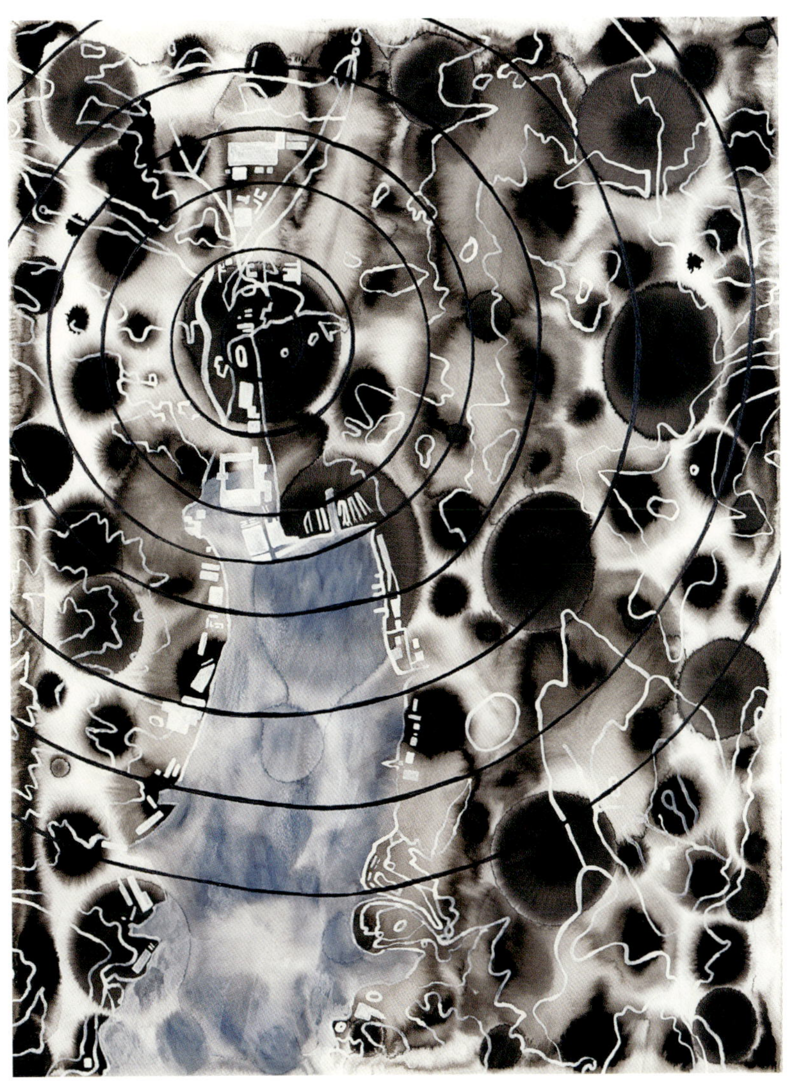

Boken Island in Enewetak Atoll, part of the Marshall Islands, is seen in this 1978 aerial photo. The lagoon at right was created by a 1956 nuclear test, which vaporized the land that used to be there. AP/Wide World Photos. Reprinted with permission.

"Dark Side of U.S. Quest for Security: Squalor on an Atoll," Howard W. French, *New York Times*, © 2001. Reprinted with permission.

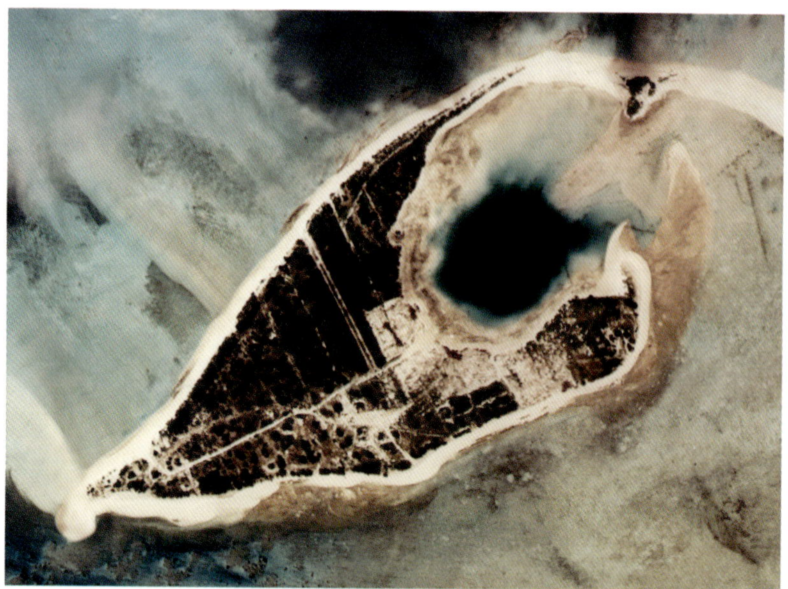

With the Bush administration vowing to develop a more expansive missile defense than ever, there is an atmosphere of excitement, at least among the Americans here, unrivaled since 1962, when a Nike Zeus rocket fired from here knocked out an incoming missile, the birth of the antimissile system era.

But the story of the Kwajalein range is more than a tale of triumphant technology. It is also an account of one of the closest things the United States has to a colonial relationship: one in which an almost idyllic small-town America sits on a sand-covered sliver of Micronesian coral, segregated from the overcrowded ghettos where native islanders have been displaced in America's pursuit of greater security.

The United States won control of the Marshall Islands' 29 atolls, comprising 1,225 low-lying coral islands, from Japan in 1944 during the bitterly fought Pacific campaign. And Washington has exercised broad control over the destiny of one of the world's smallest nations ever since.

After the war, the United States administered the territory under a United Nations mandate until 1986, when the country entered into a "free association" with Washington. Since then, the United States has secured use of its base here by a series of 15-year renewable agreements.

Those pacts make the United States responsible for the country's

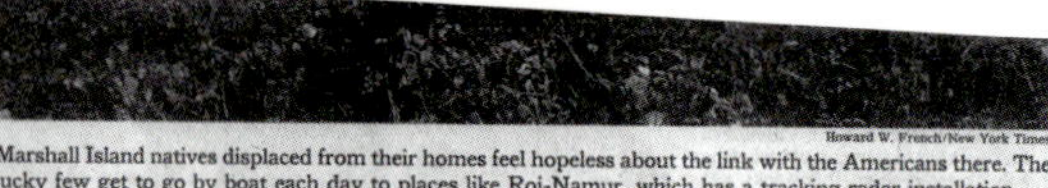

Howard W. French/New York Times

Marshall Island natives displaced from their homes feel hopeless about the link with the Americans there. The lucky few get to go by boat each day to places like Roi-Namur, which has a tracking radar installation.

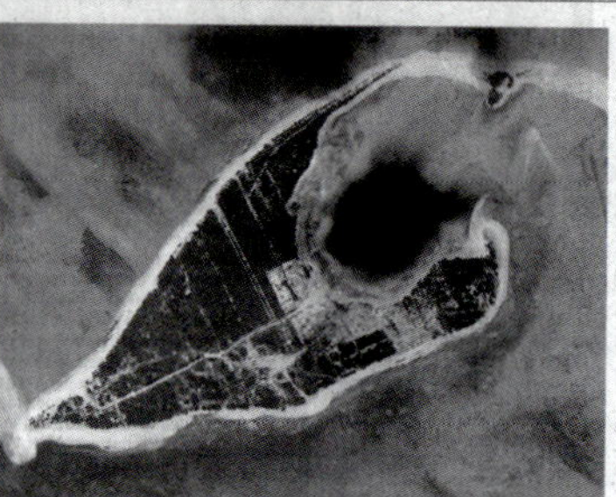

Associated Press

A 1978 photo shows the crater from an atomic test at Boken Island. Americans still shape the destiny of places like Kwajalein Atoll.

square mile island is now home to nearly a third of the Marshall Islands' total population.

Speaking warily, a woman who works at the Cafe Pacific, a dining hall that serves hearty breakfasts of pancakes, grits, sausages and eggs to the American workers on Kwajalein, described the mood of the Marshallese natives who live on Ebeye and the other islands as one of hopelessness and defeat.

"There used to be protests about the situation here," said the cafeteria worker. "But the Americans control the biggest island in this atoll, and they decide who gets jobs, and how much we get paid, too. We know there is something wrong here, but we feel like a mouse up against an elephant. What can we do?"

Many Americans blame Marshallese outright for their plight, saying they have wasted large sums of aid and that they crowd together in substandard housing because of the islanders' communal culture.

"Everyone knows that Ebeye is

Boken Island, Republic of the Marshall Islands, 1945–2001

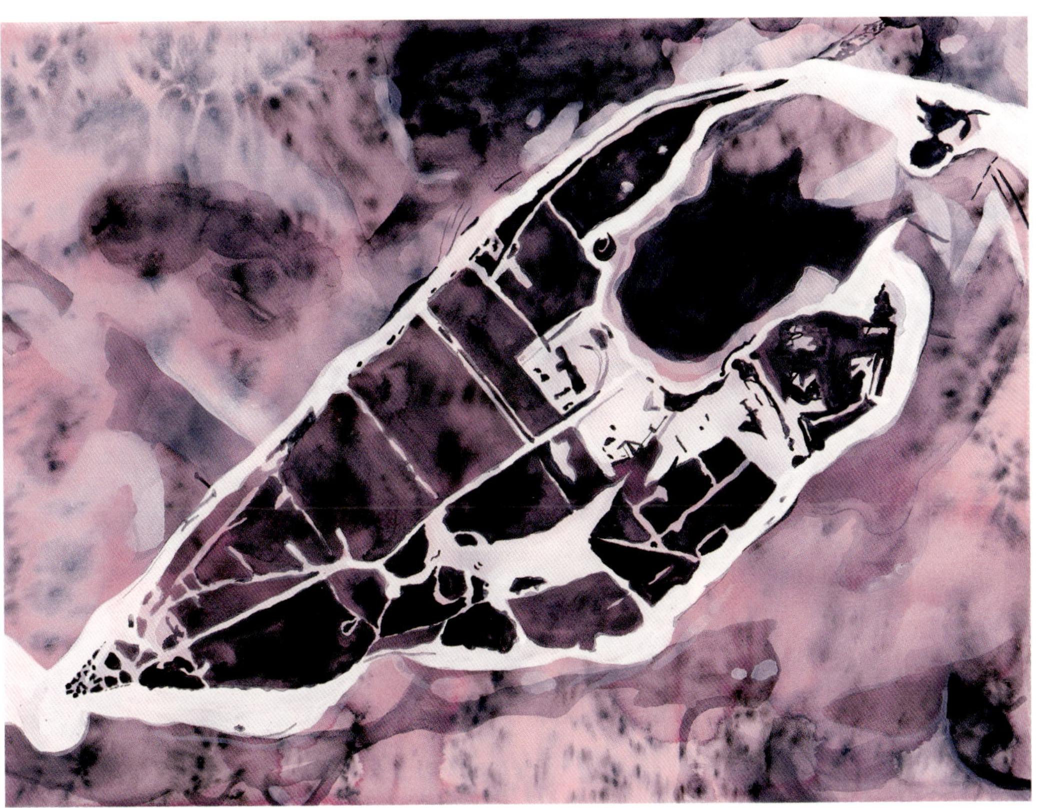

Vienna, Austria, 1945

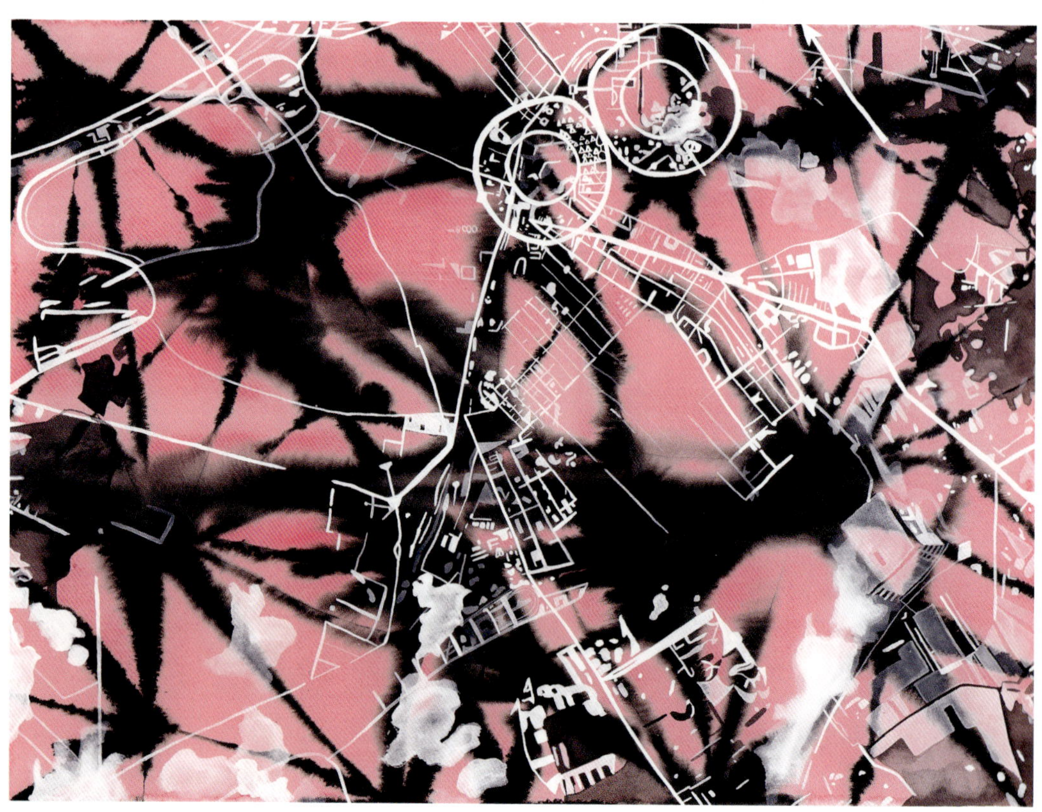

The Firebombing of Tokyo, Japan, 1945

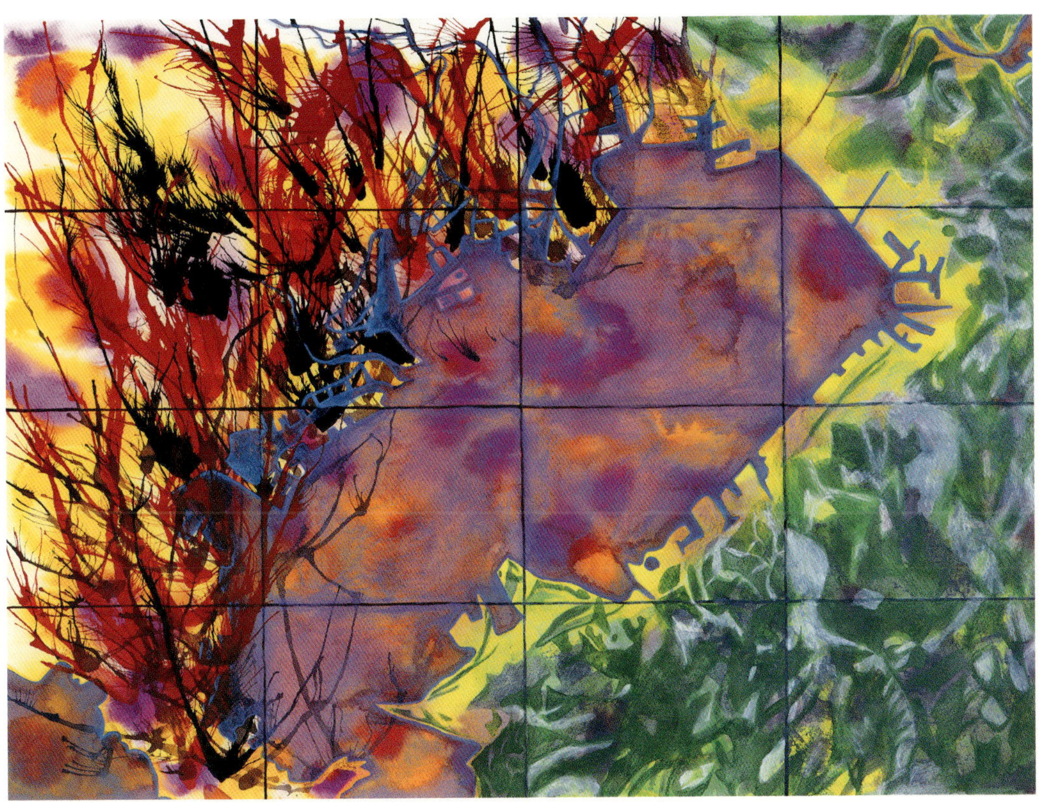

Bikini Atoll, Republic of the Marshall Islands, 1946–1958

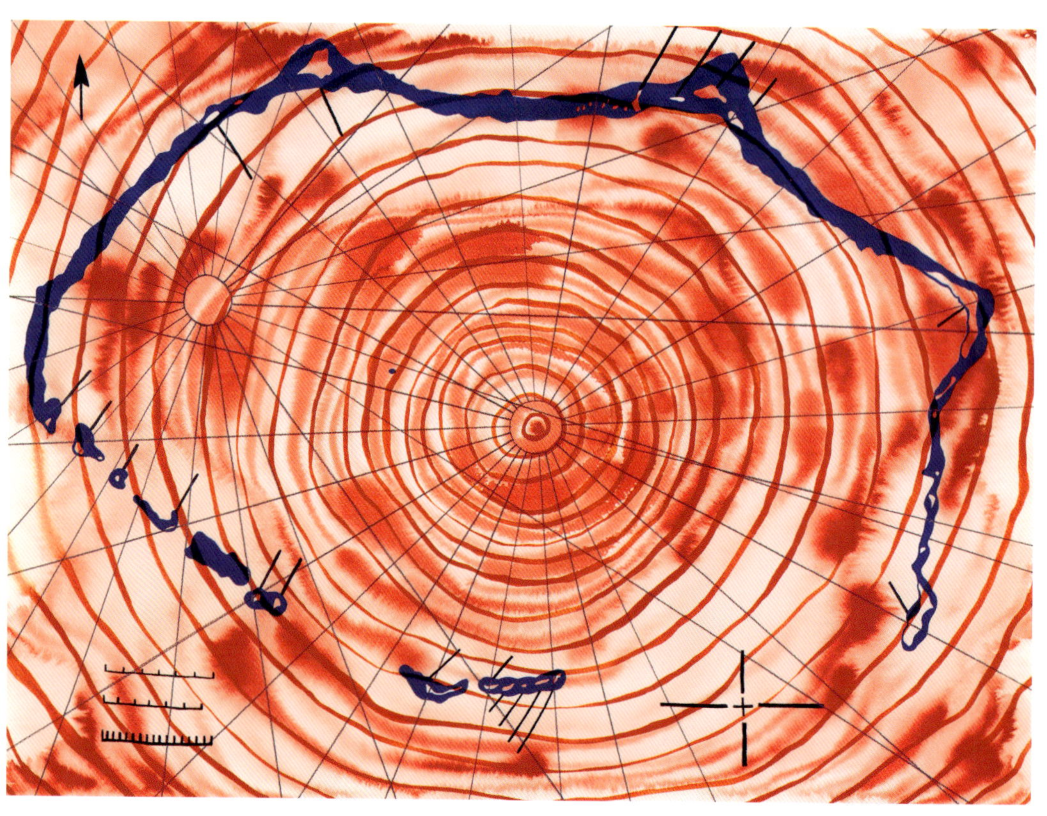

Enewetak Atoll, Republic of the Marshall Islands, 1948–1958

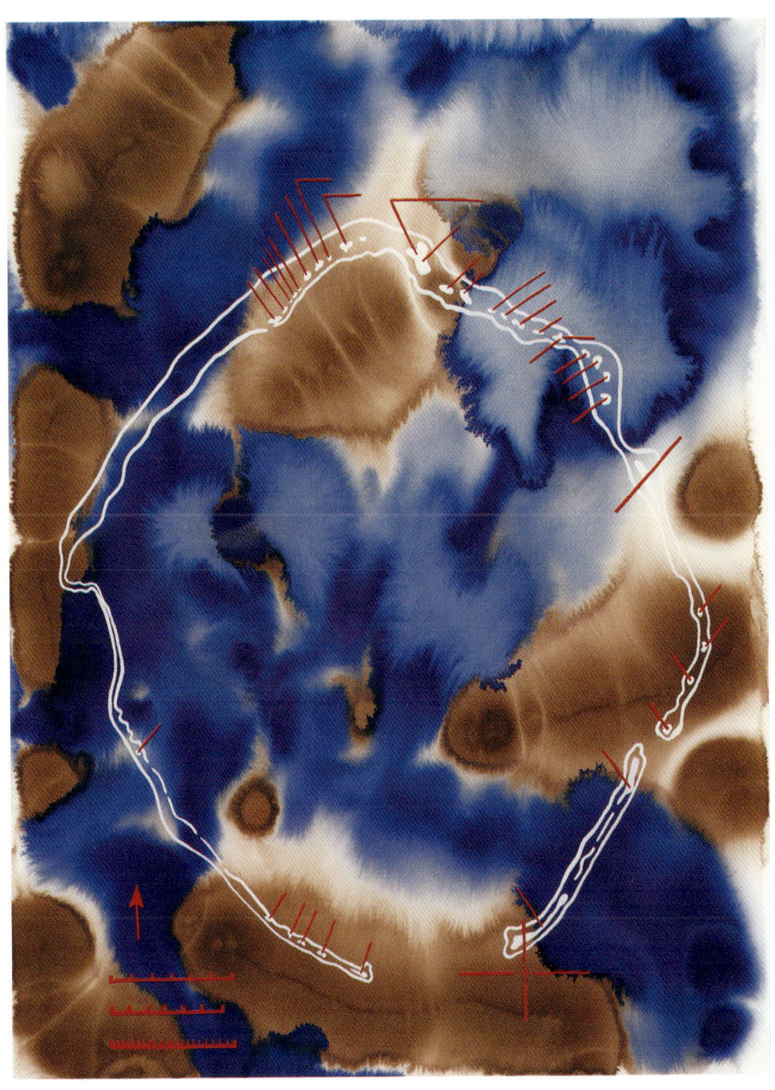

Korea, 1950–1953

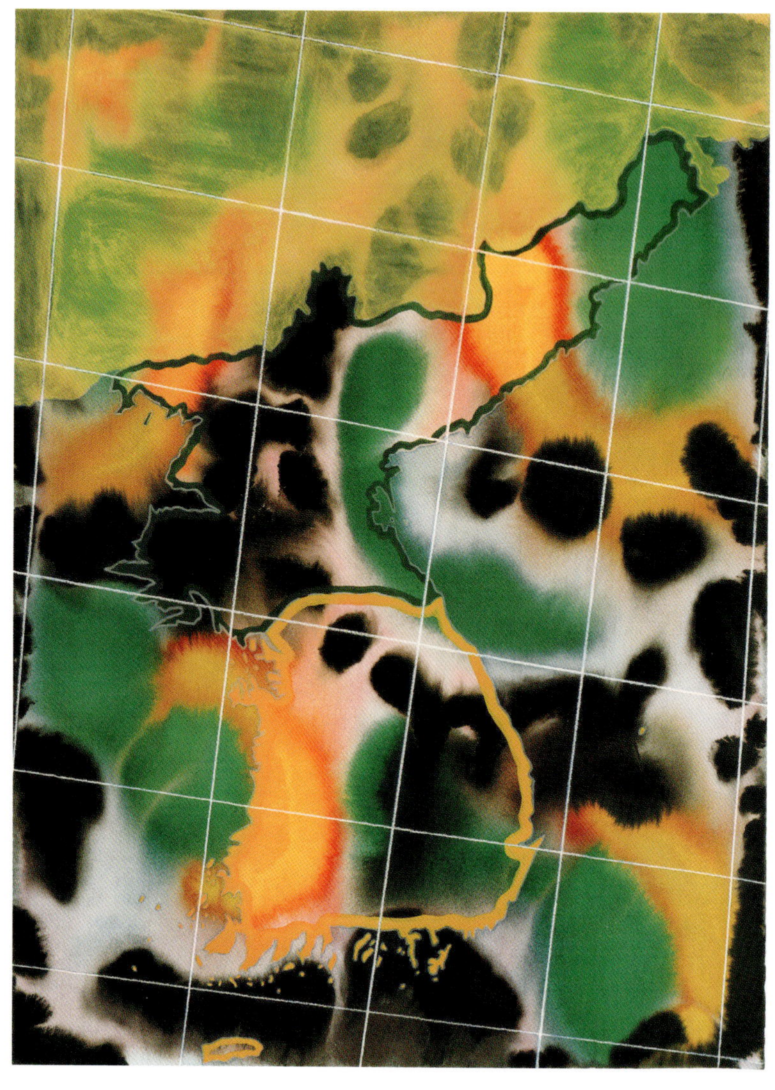

Proving Ground, Nevada I, U.S., 1951–1992

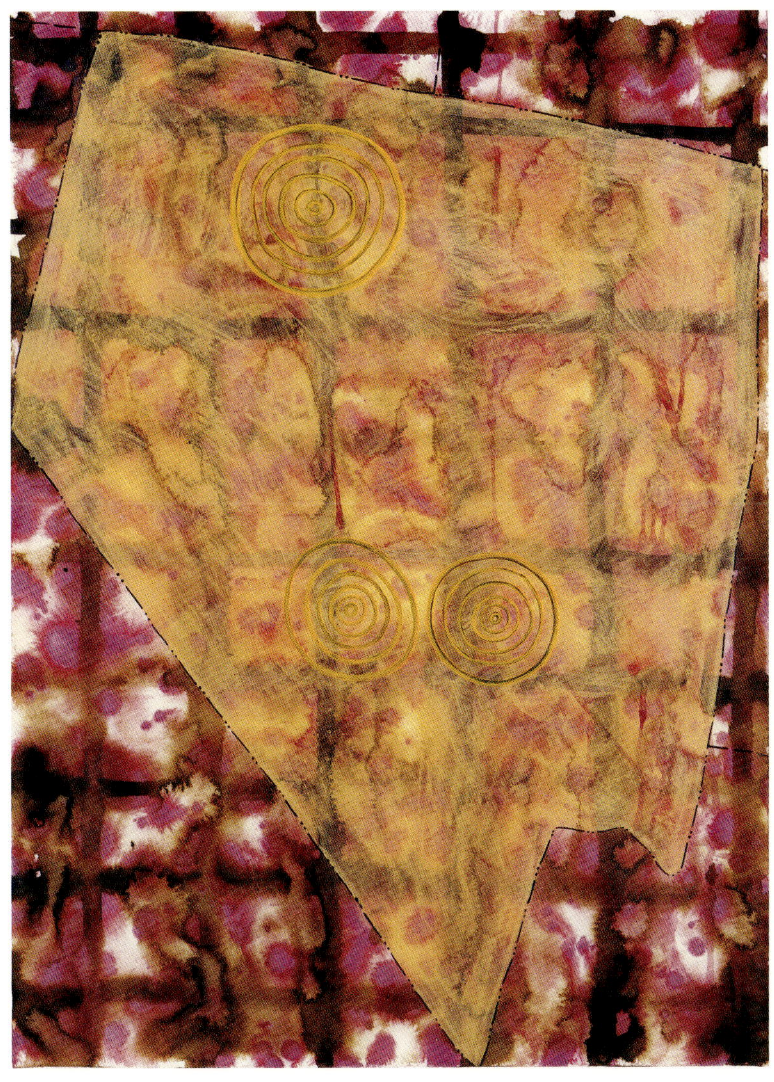

Guatemala, 1953–1980

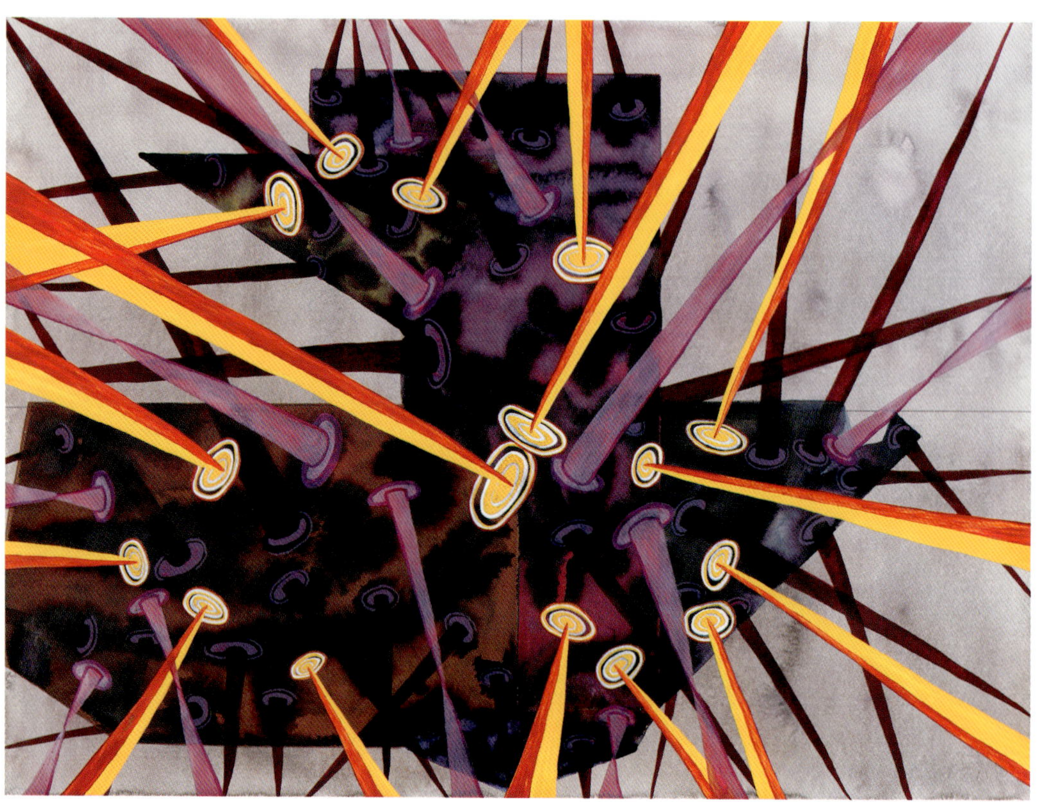

Iran, 1953 and 1987–1988

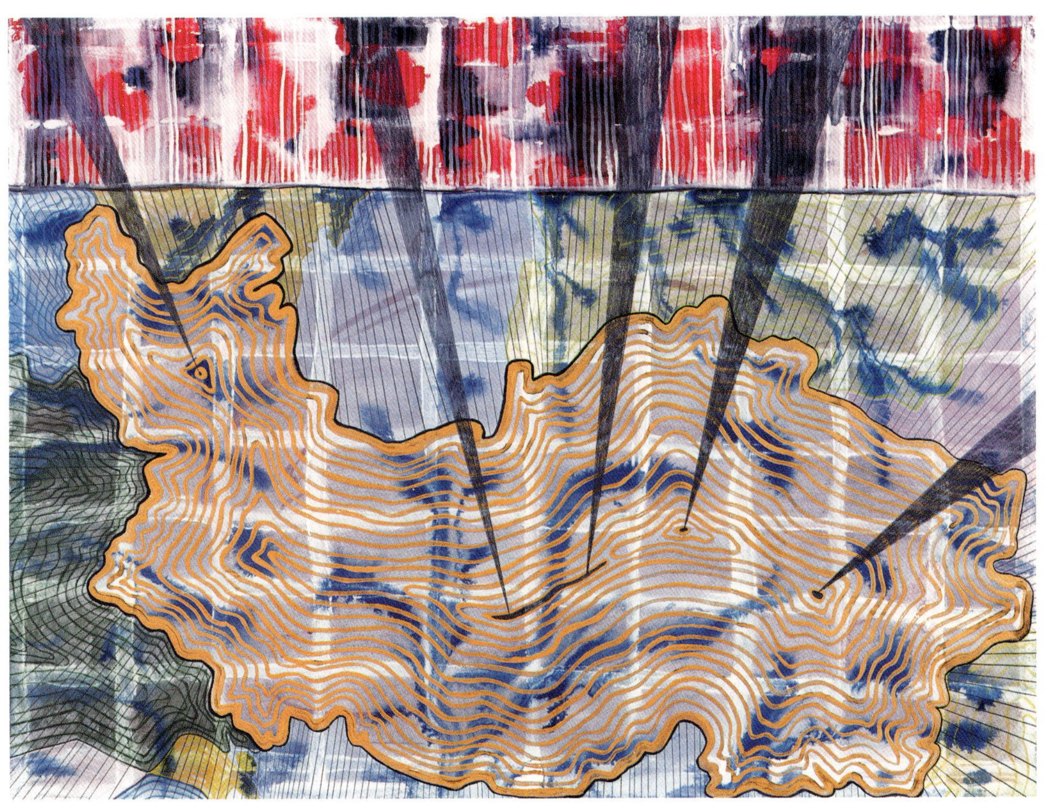

Scott Kirsch, "Peaceful Nuclear Explosions and the Geography of Scientific Authority," in *Professional Geographer*, Blackwell Publishing, 2000. Reprinted with permission.

Peaceful Nuclear Explosions **189**

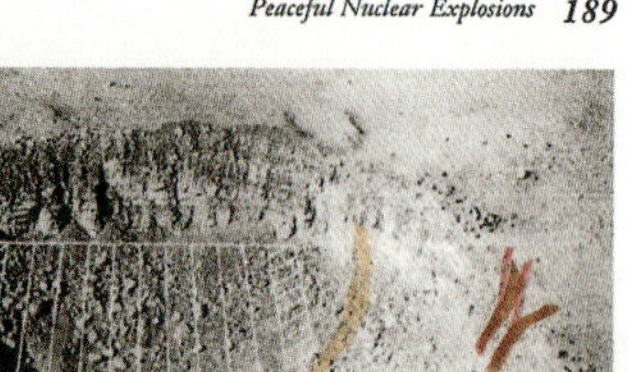

Figure 2: *Schooner crater and football field. Courtesy Office of History and Historical Records, Lawrence Livermore National Laboratory.*

Schooner was the last of six Plowshare nuclear cratering tests conducted at the Test Site during the 1960s, though Livermore personnel continued to plan additional experiments and to scan the globe for sites appropriate for nuclear excavation.

In December, 1970, when the Army Corps of Engineers reluctantly rejected plans for a new sea level canal to be blasted across Central America with some 250 nuclear explosives, Congressional funding was curtailed and Plowshare research on the once promising idea of nuclear excavation effectively ceased. Despite some success convincing the physics and engineering communities and various government and commercial interests that nuclear excavation was a viable technology, Livermore was ultimately unable to take its geographical engineering beyond the borders of the Nevada Test Site.[18] If boundaries were used as resources by the Plowshare scientists, then they were socially constructed *limits* as well, geographical but not absolute, to the authority to develop "peaceful uses" of nucle[illegible]plosives.■

Notes

[1] This [illegible]rch is based primarily on data collected fro[illegible] Office of History and Historical Records at the Lawrence Livermore National Laboratory (hereafter LLNL), Livermore, CA, in October 1995 and February 1997, and from the U.S. Department of Energy Coordination and Information Center, Las Vegas, NV (hereafter CIC) in June–July 1996.

[2] I use the phrase "geography of science" rather than Livingstone's "historical geography of scientific ideas" here in order to emphasize, first, the applicability of contemporary as well as historical research to this project, and second, though it is certainly implicit in Livingstone (1995), that I am more concerned with science "as practice and culture" than in terms of a history of ideas (see Pickering 1992).

[3] The point of this position is not to collapse science into a radical relativism in which any one scientific finding is as good as another, nor to reduce scientific truth to a mere exercise in power relations. Nor should it imply that scientific constructions of nature are entirely and irreducibly social. Rather, it opens a theoretical space for evaluating the historically and geographically contingent nature of scientific facts in which, as Demeritt (1998, 178) puts it, "The criterion for success for scientific theory is empirical adequacy and pragmatic achievement, not ultimate truth or falsity."

[4] For a historical analysis of the changing orientations of basic research in post-World War II American science, see Forman (1987).

[5] As I was reminded by a former Livermore scientist during a 1999 talk that I gave at the University of Kansas Geography Department, much of this work—especially that related to the nuclear cratering experiments—also fulfilled military data needs concerning the effects of nuclear weapons. Reciprocally,

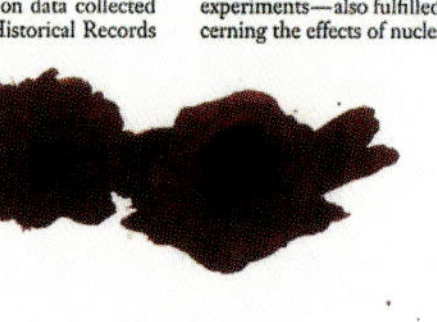

Nevada Test Site II, U.S., 1955–1968

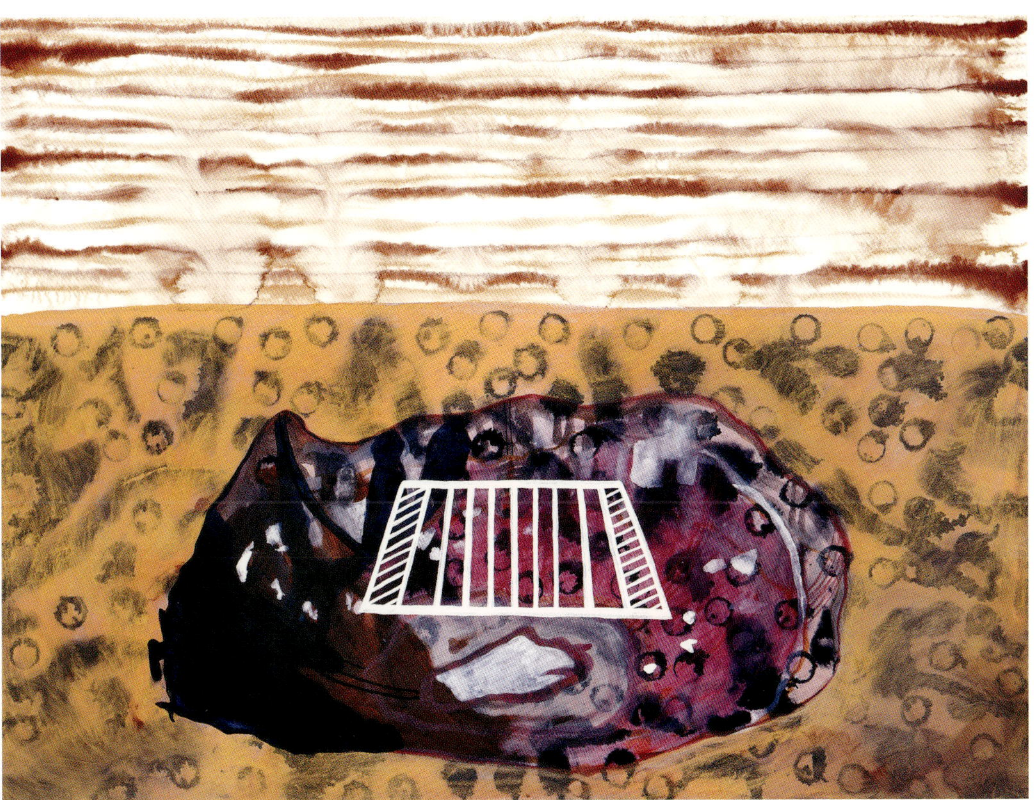

Johnston Atoll, U.S., 1958–1962

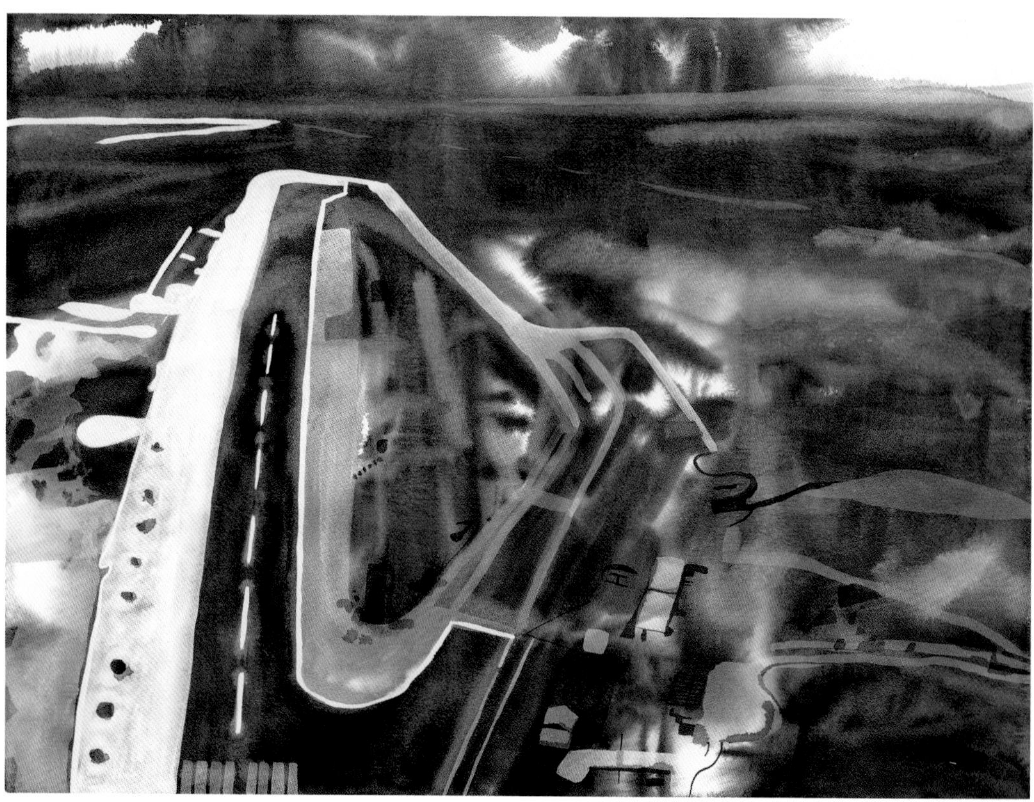

Haiti, 1959

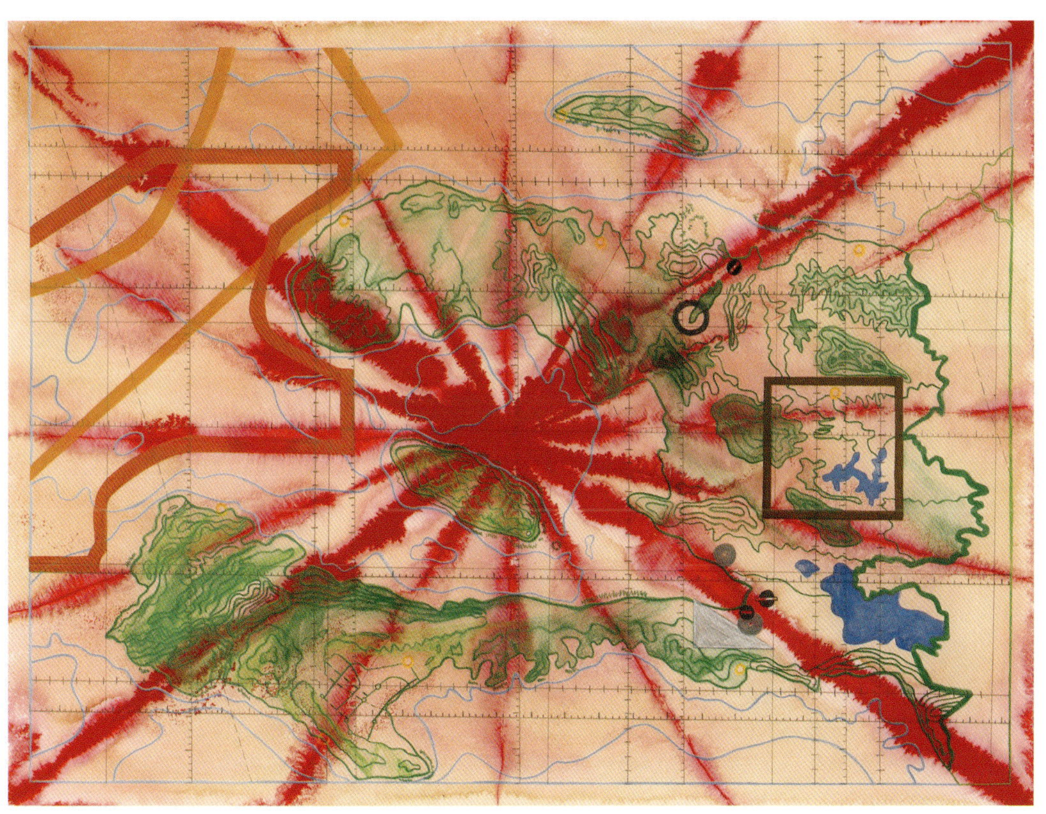

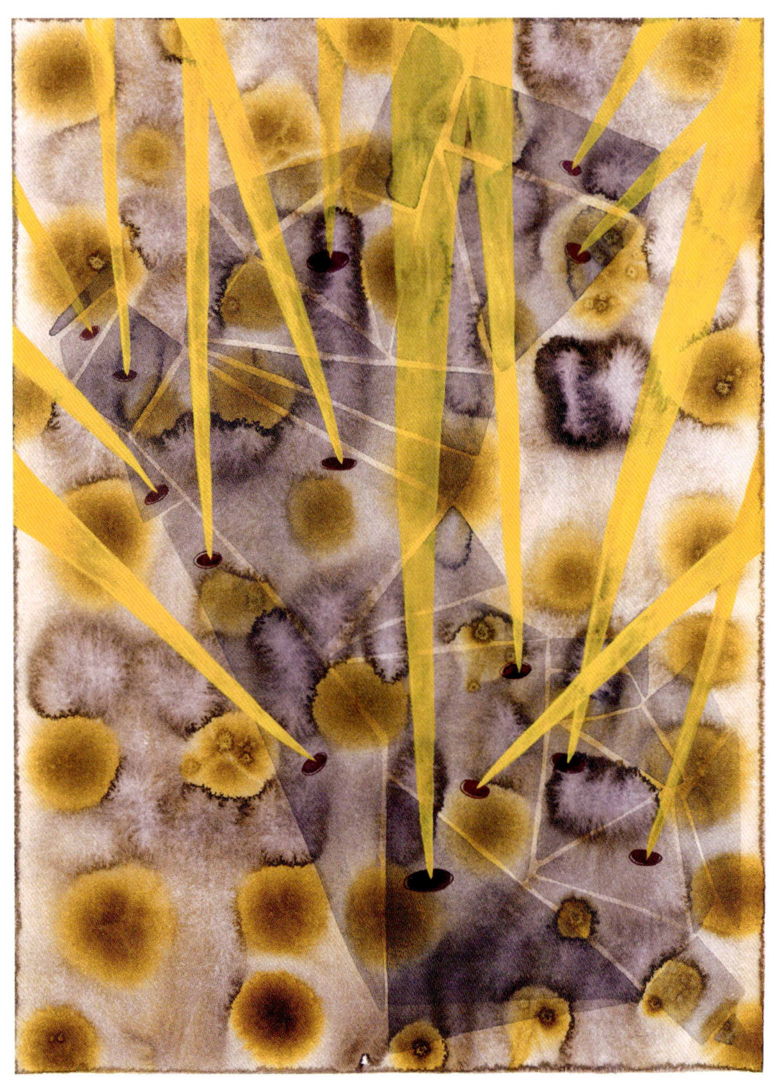

The Belgian Congo, Zaire, now the Democratic Republic of the Congo, 1960–1964

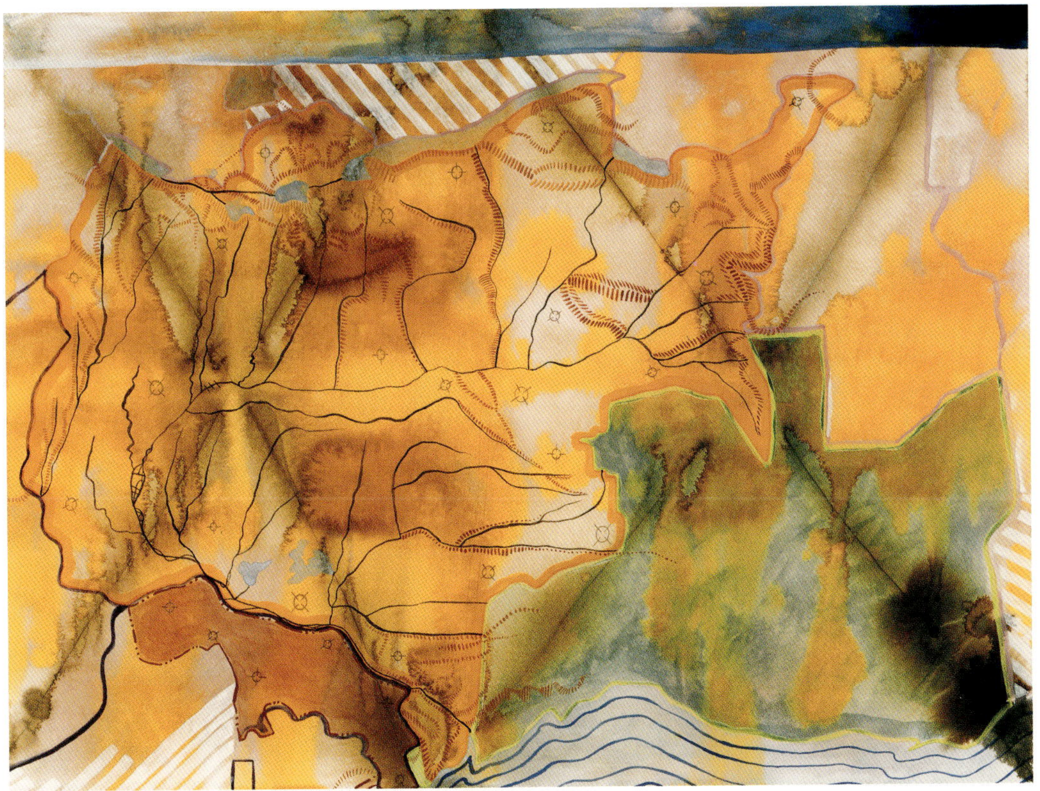

Reference Map of Vietnam

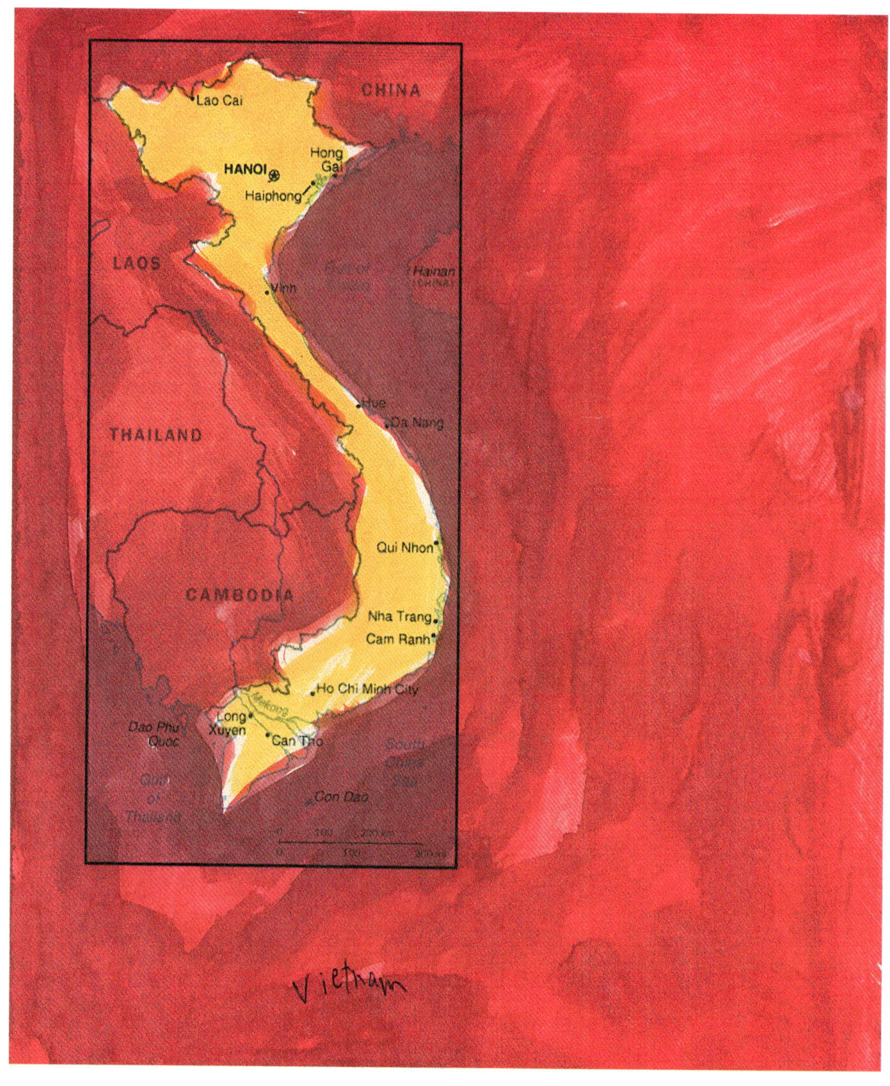

Vietnam, 1961–1971

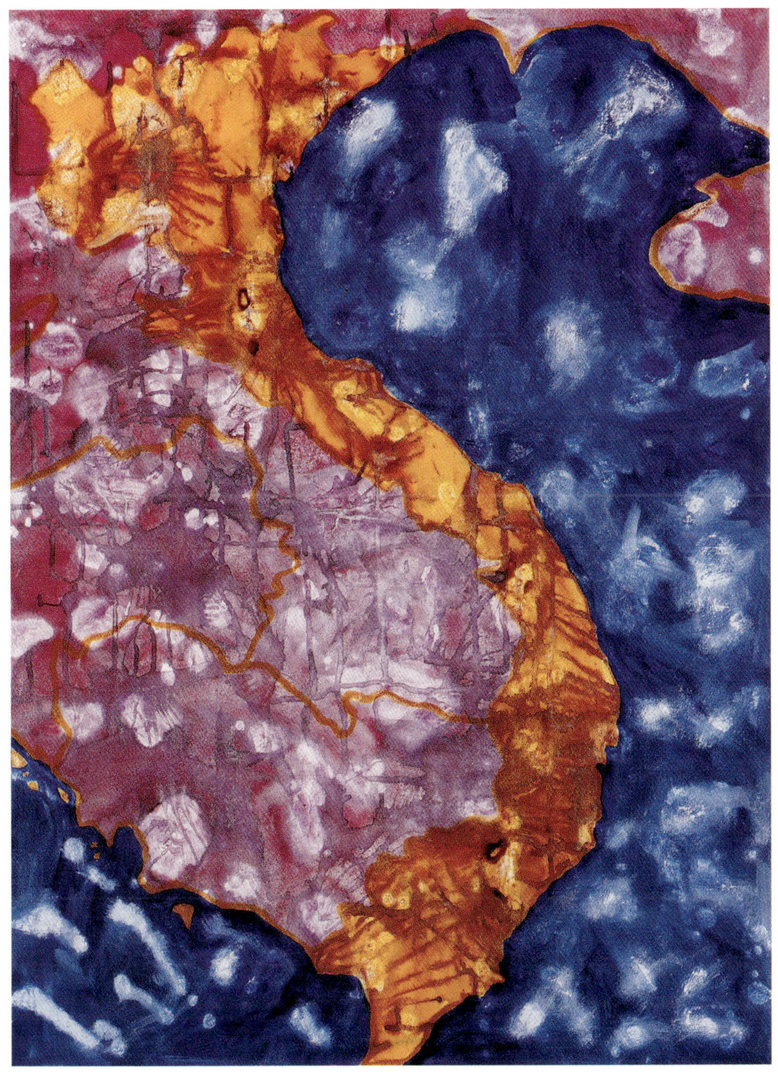

Kiritimati or Christmas Island, Republic of Kiribati, 1962

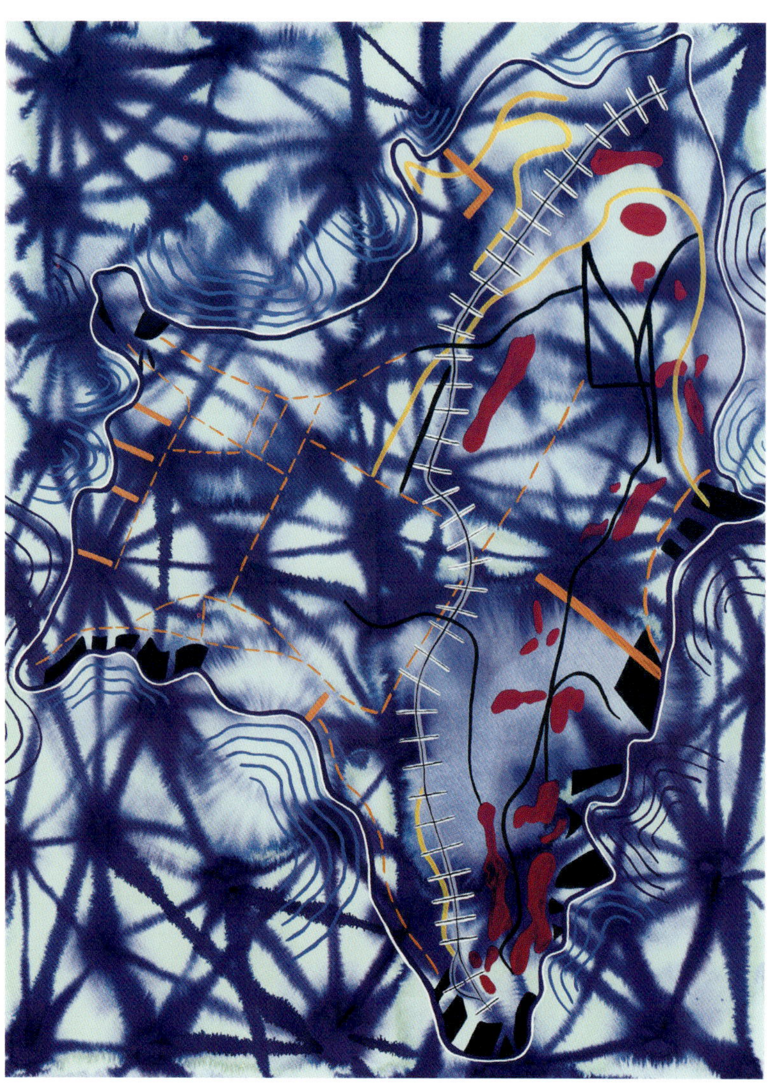

Laos, 1964–1973

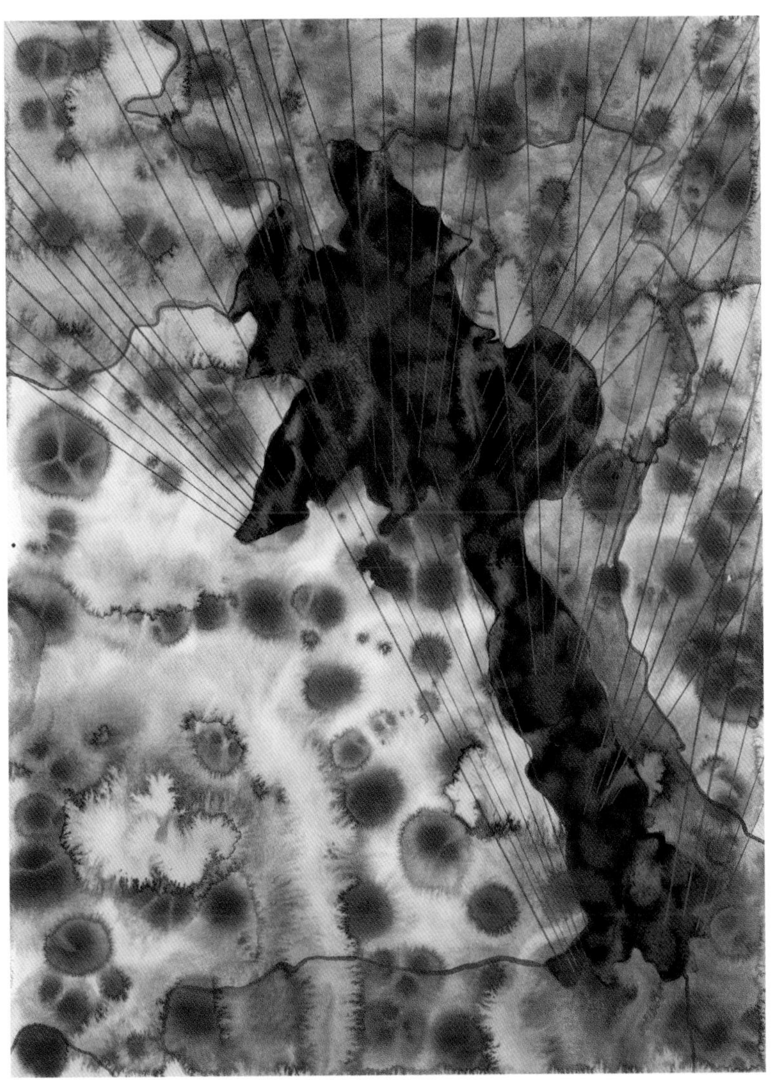

Mississippi, U.S., 1964–1970

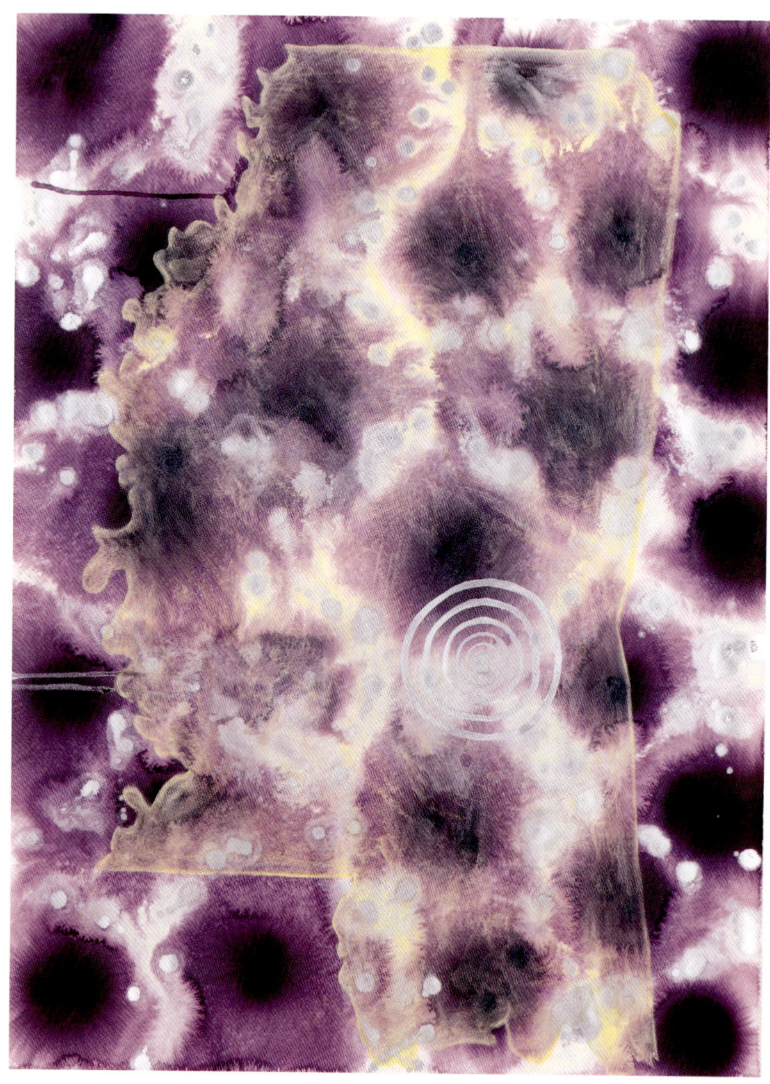

Amchitka Island, Alaska, U.S., 1965–1971

following pages
Cambodia, 1969–1973

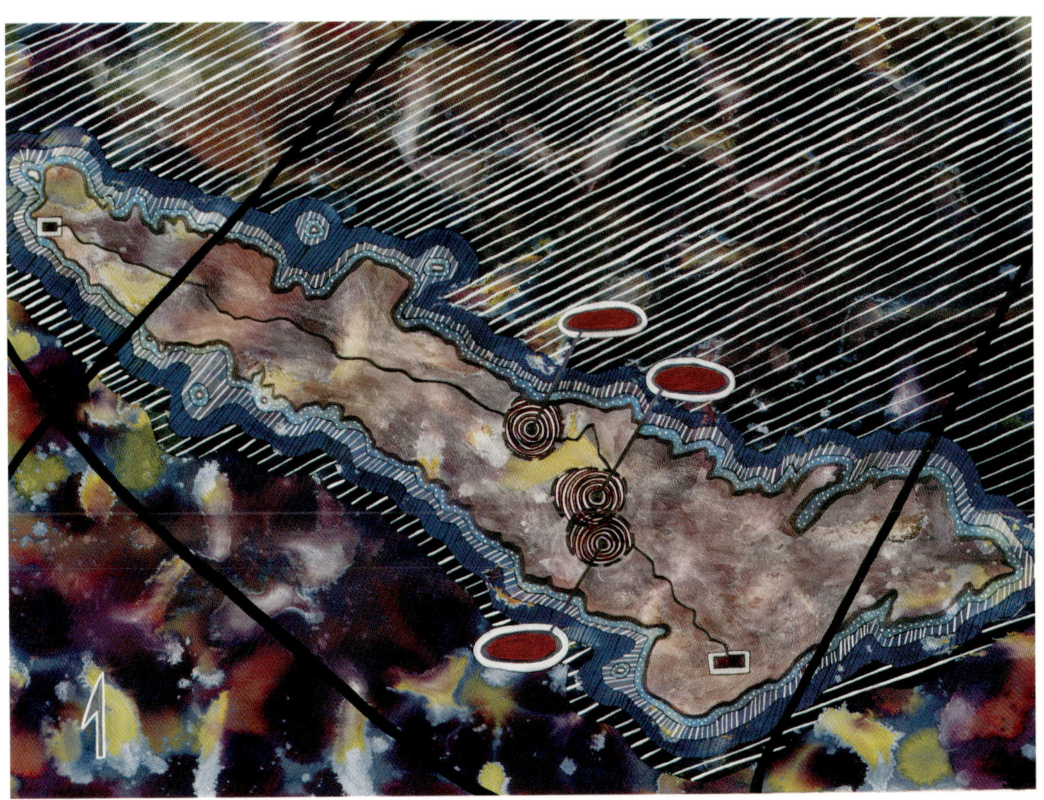

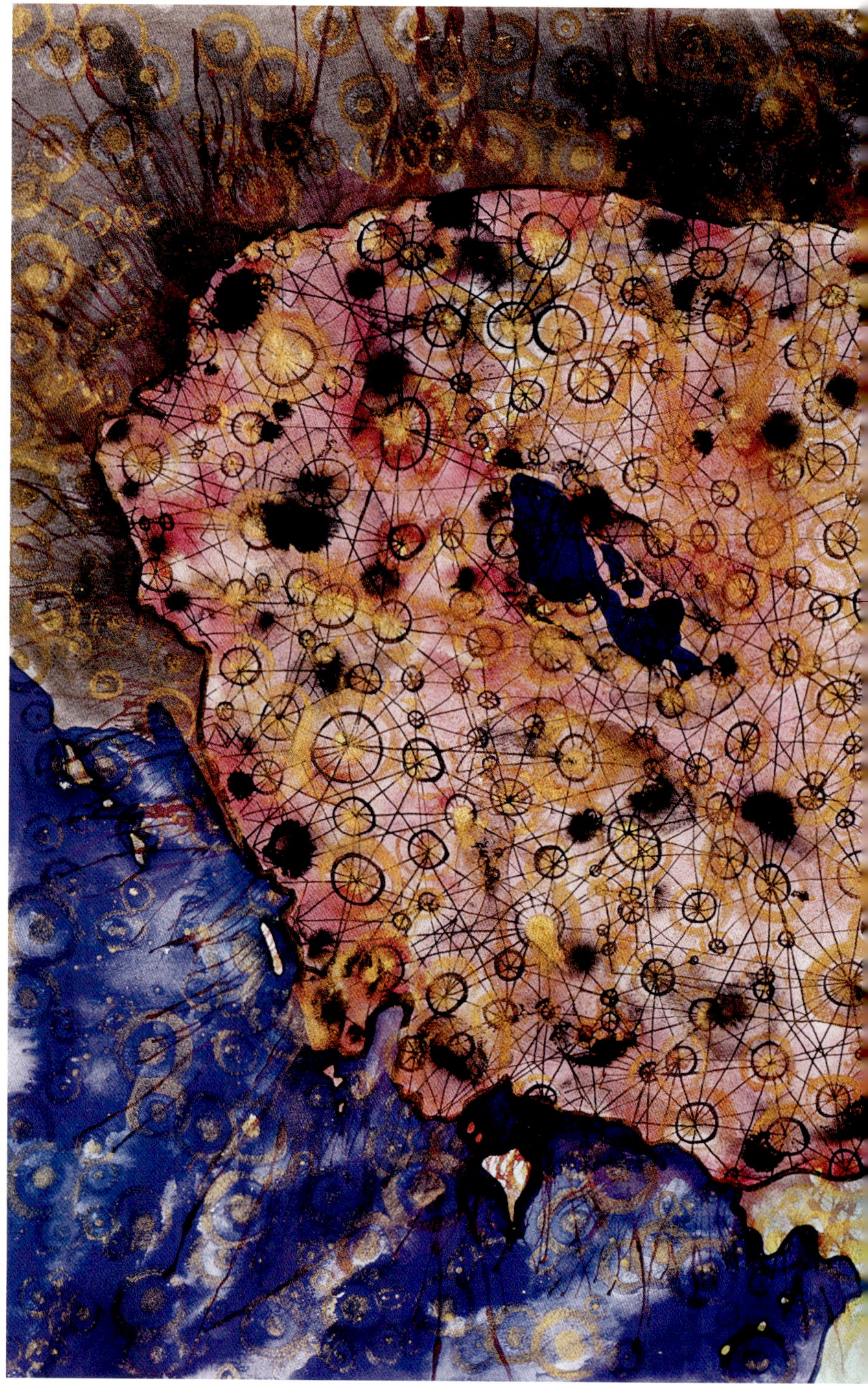

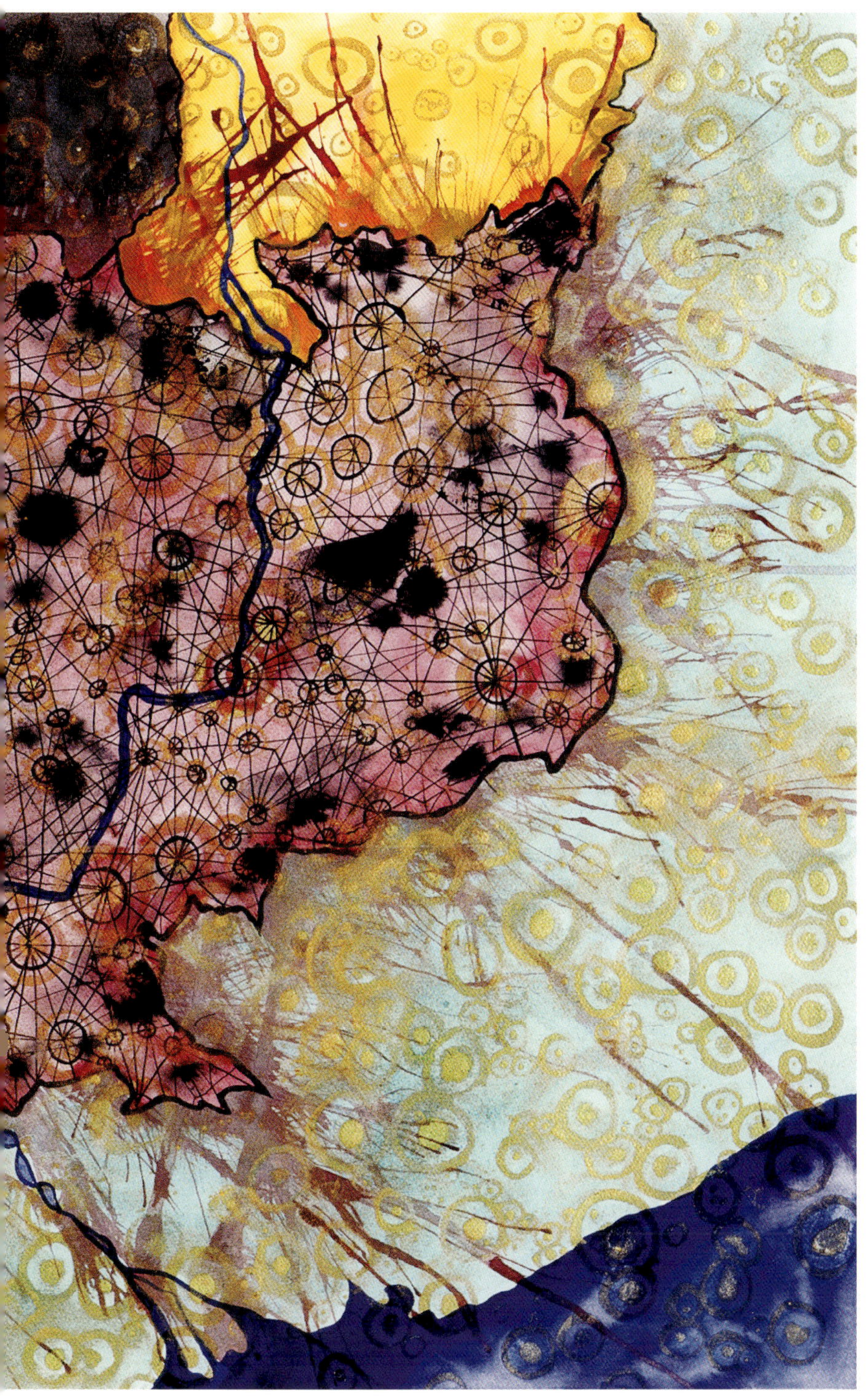

Afghanistan I, 1979 and Infinite Reach, 1998

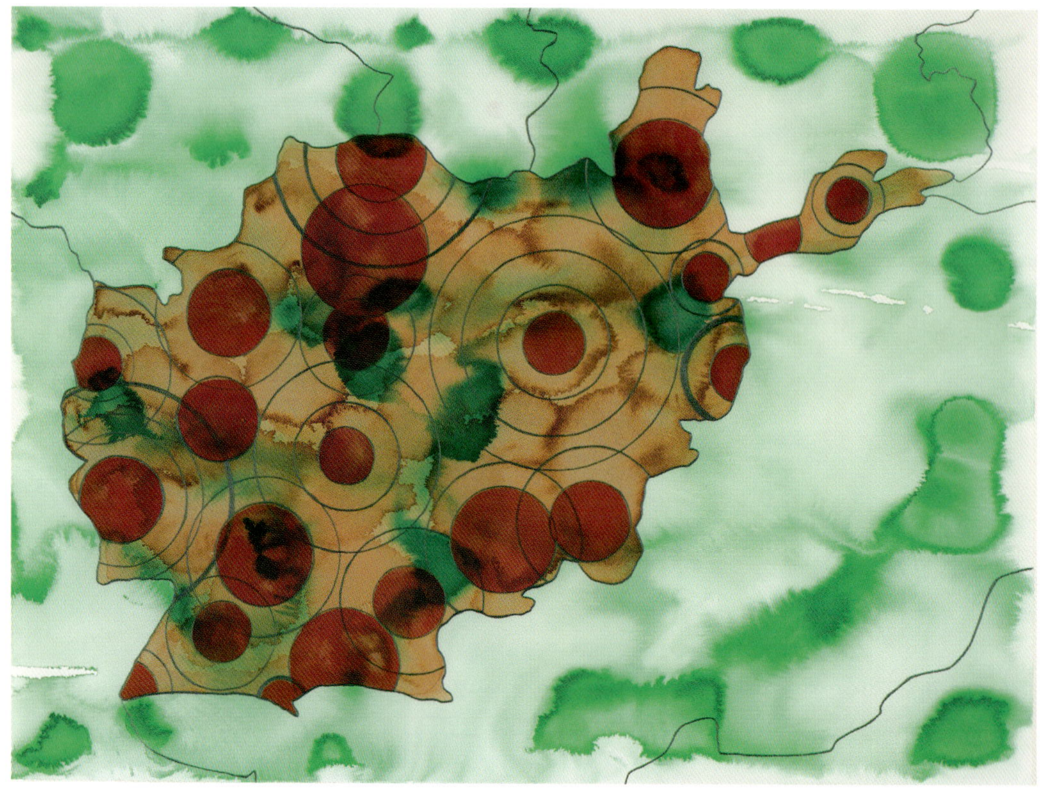

El Salvador, 1980–1994

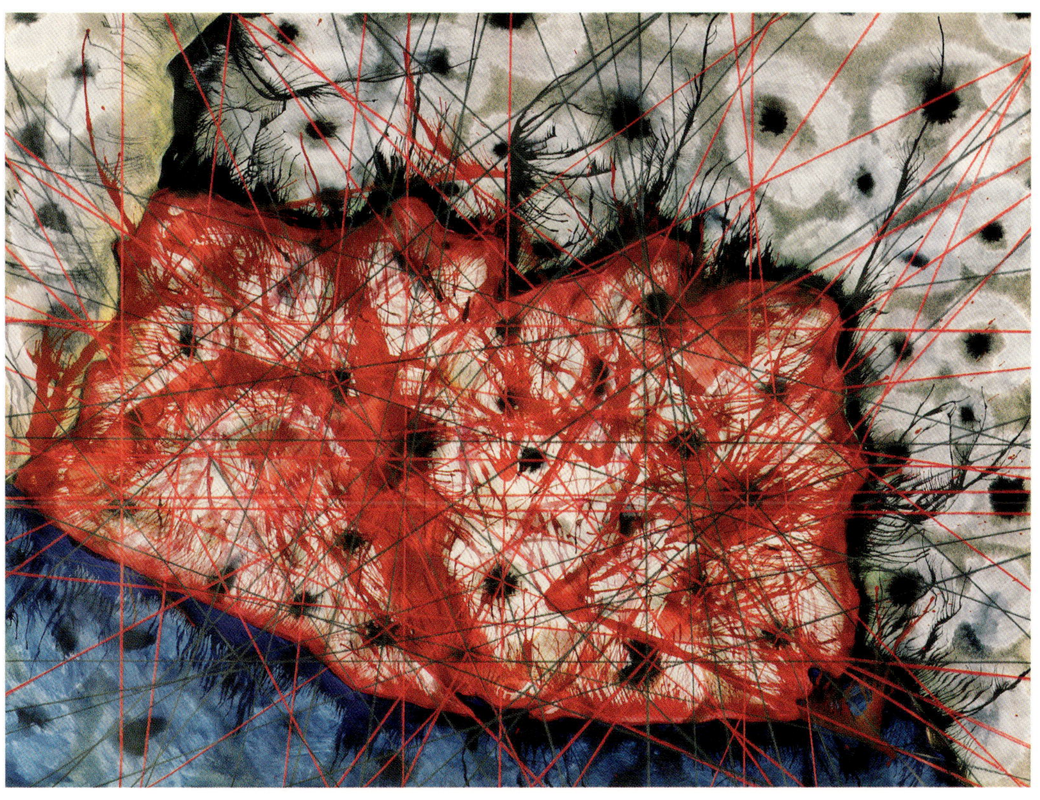

Lebanon, 1983–1984 and 2006

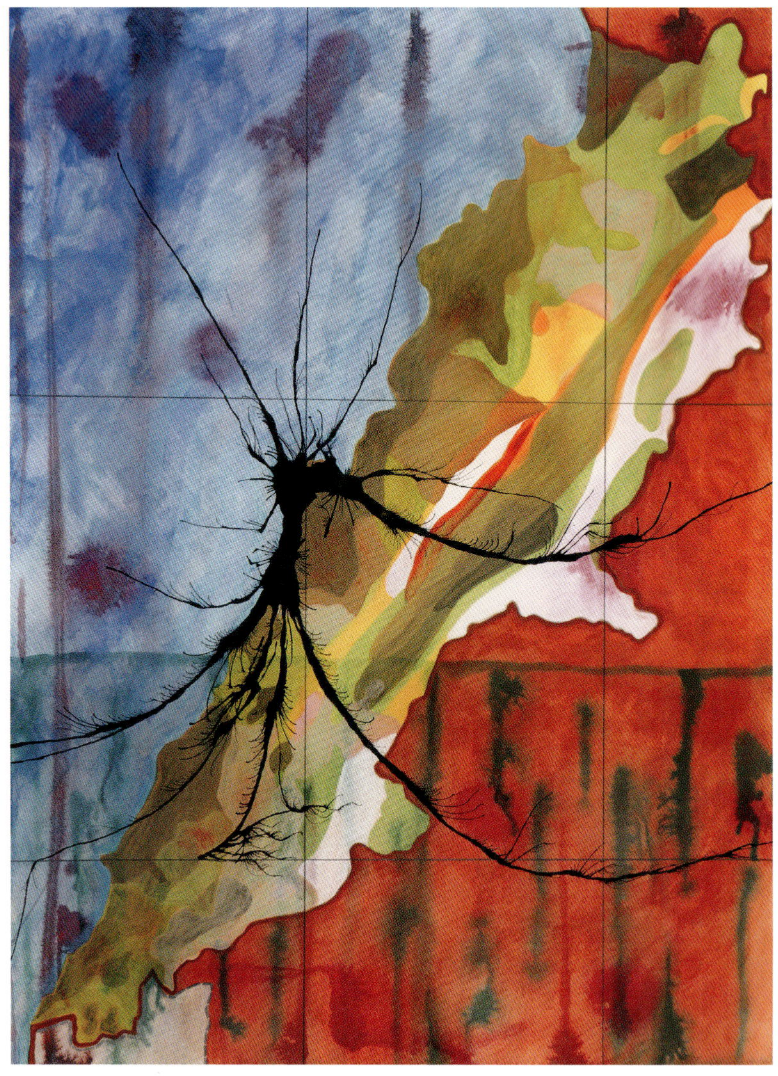

Grenada, Operation Urgent Fury, 1983

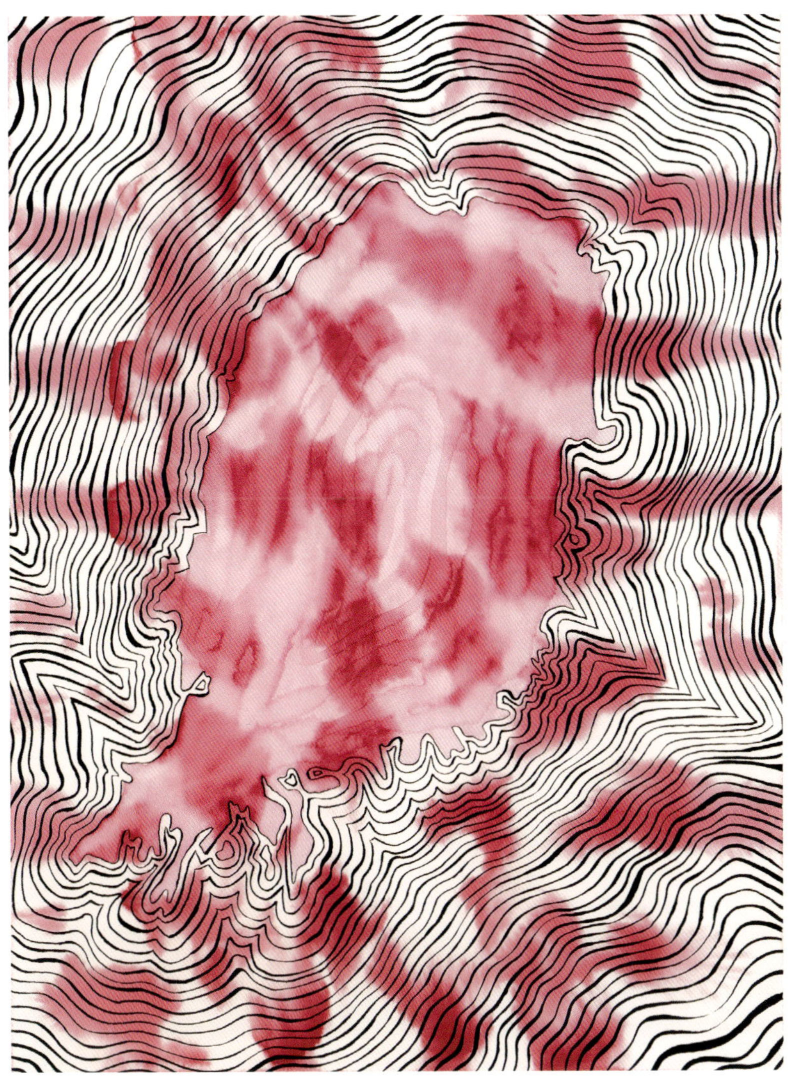

Reference Image of Libya

IMINT - Libya http://www.fas.org/irp/imint/lib_b1.htm

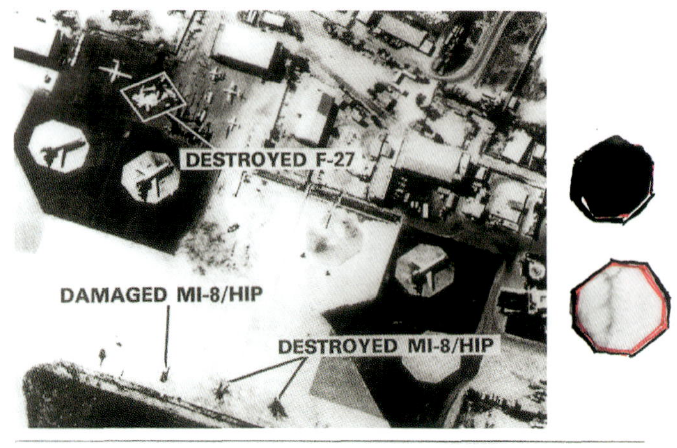

FAS | Intelligence | IMINT Gallery | Libya | Index | Search | Join FAS

Federation of American Scientists

http://www.fas.org/irp/imint/lib_b1.htm

Updated Sunday, May 12, 1996 - 6:19:10 AM

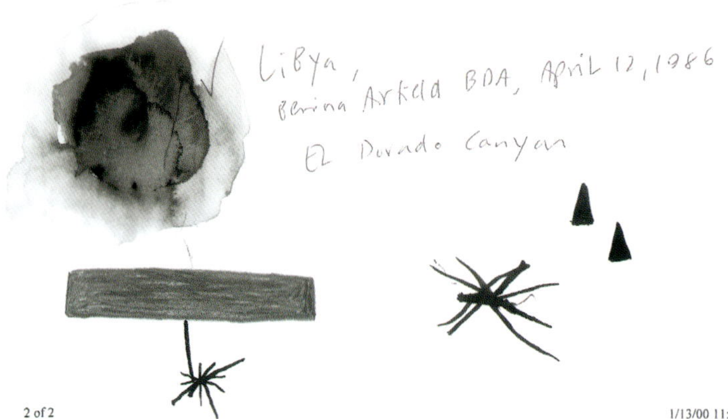

2 of 2 1/13/00 11:44 AM

Libya, Operation El Dorado Canyon, 1986

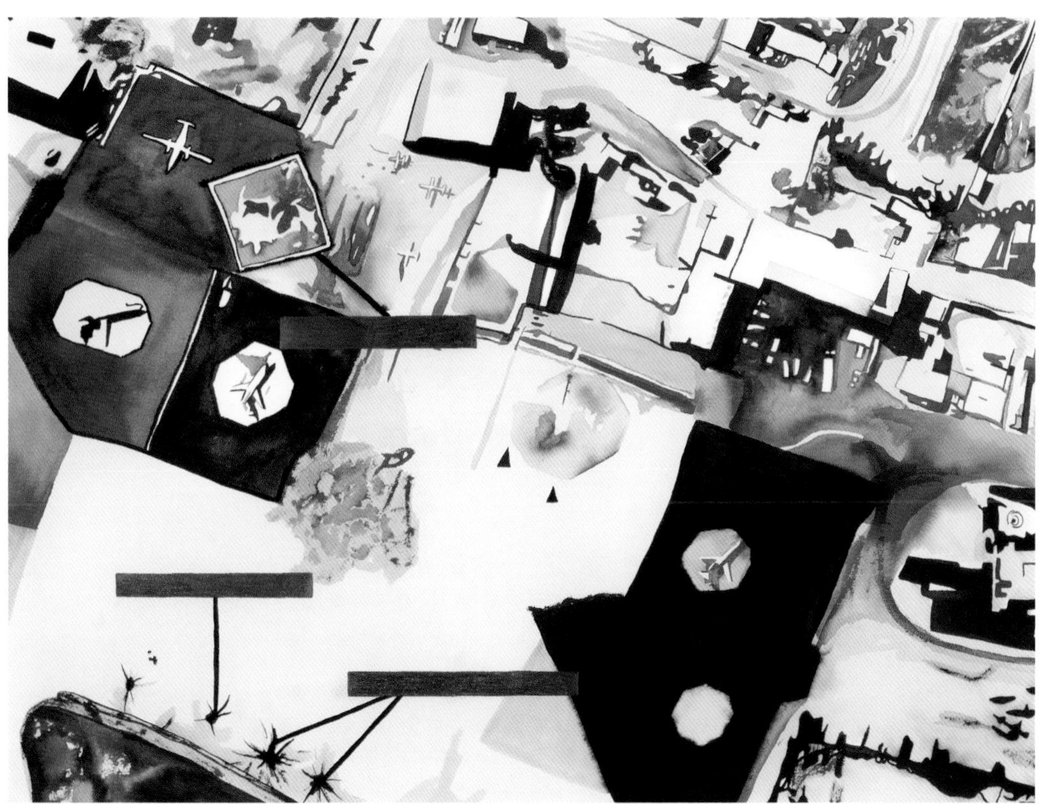

Philadelphia, the Firebombing of M.O.V.E., U.S., 1985

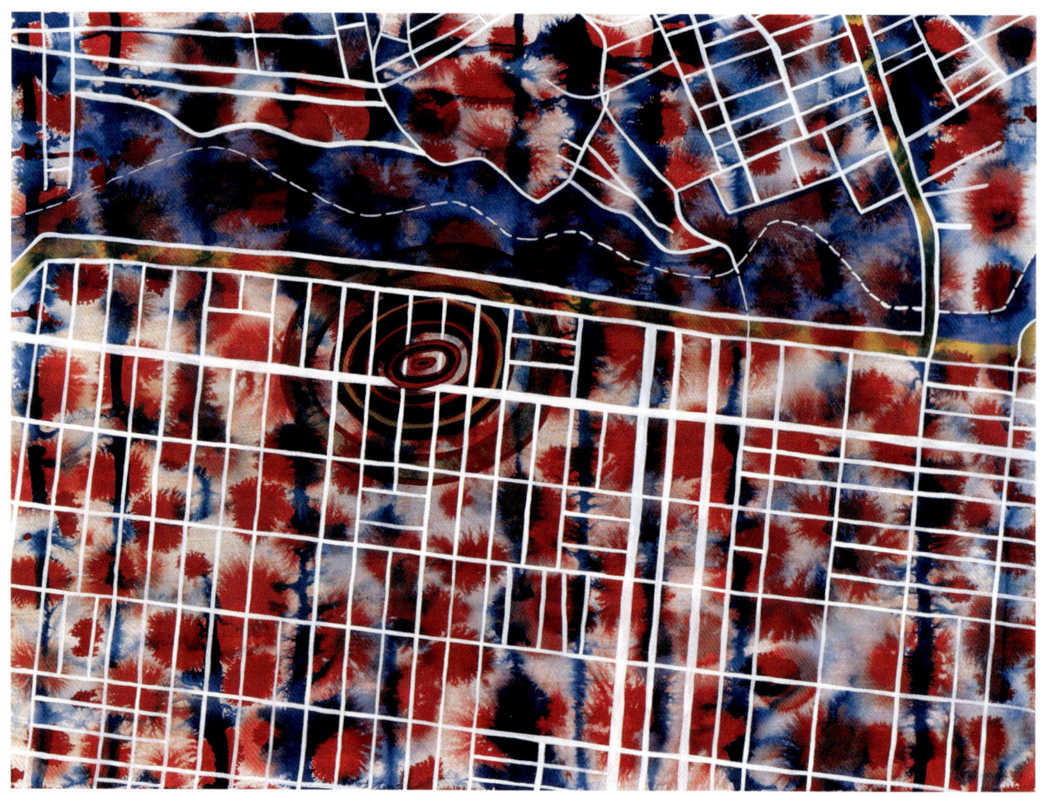

Panama, Operation Just Cause, 1989–1990

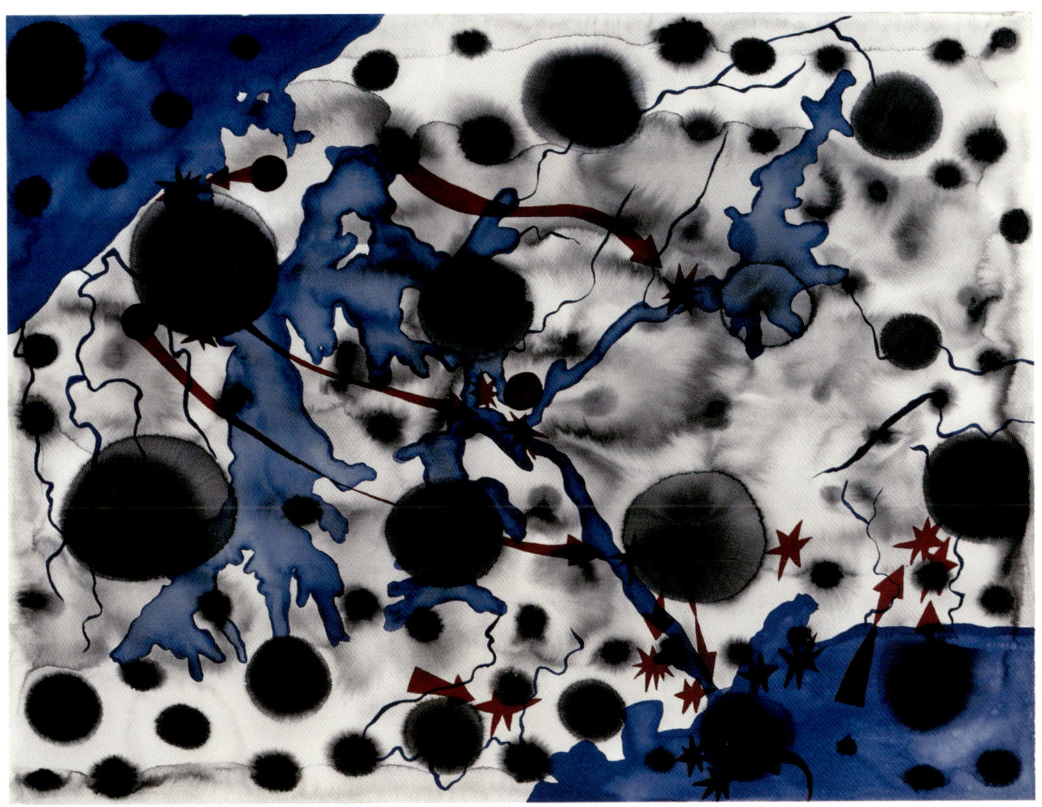

Baghdad, Iraq, 1990–Ongoing

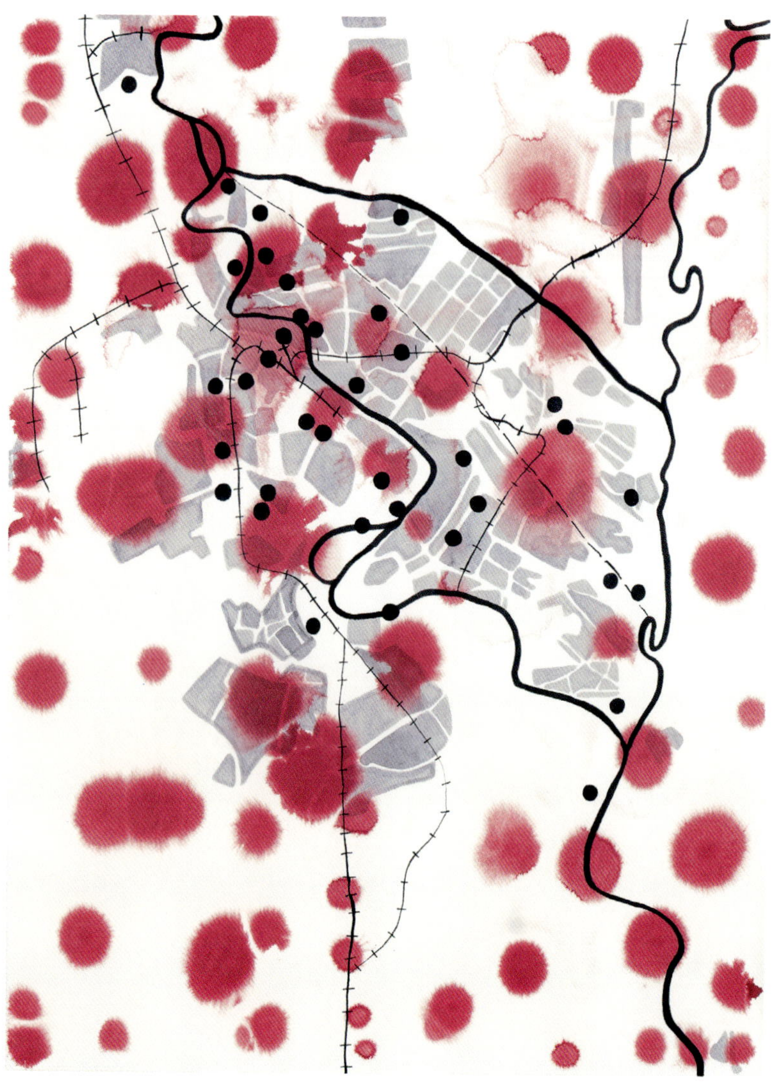

Kuwait, 1991

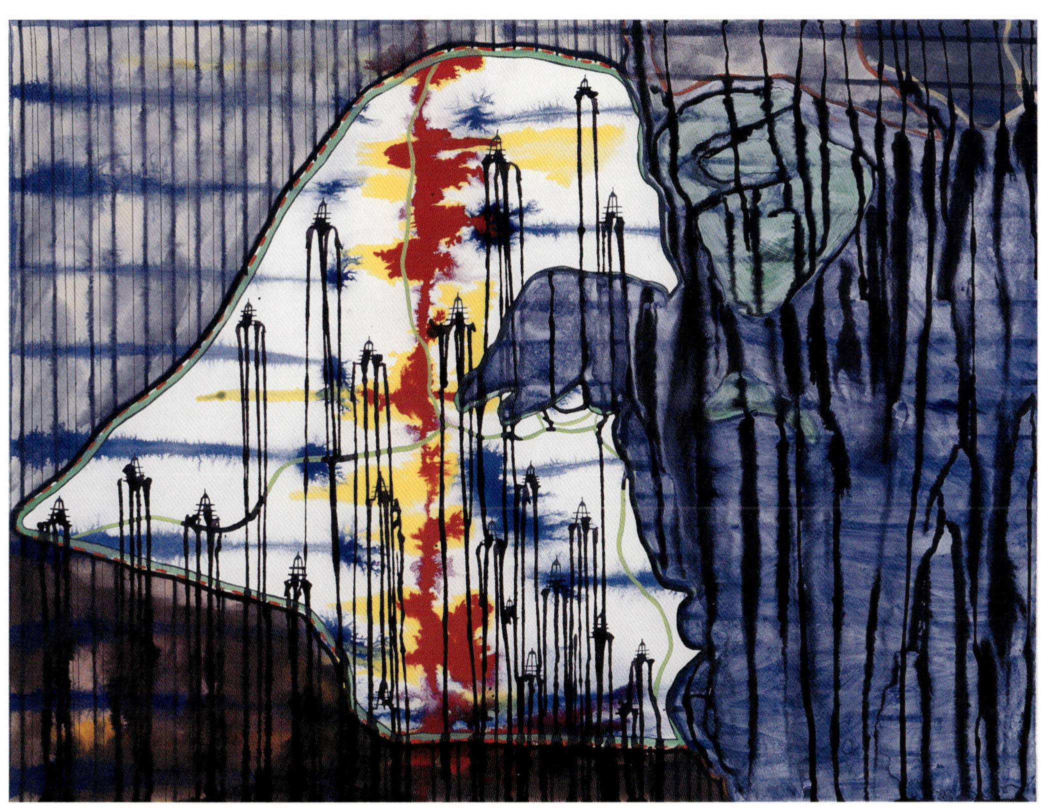

Shifa Pharmaceutical Plant, Khartoum, Sudan, 1998

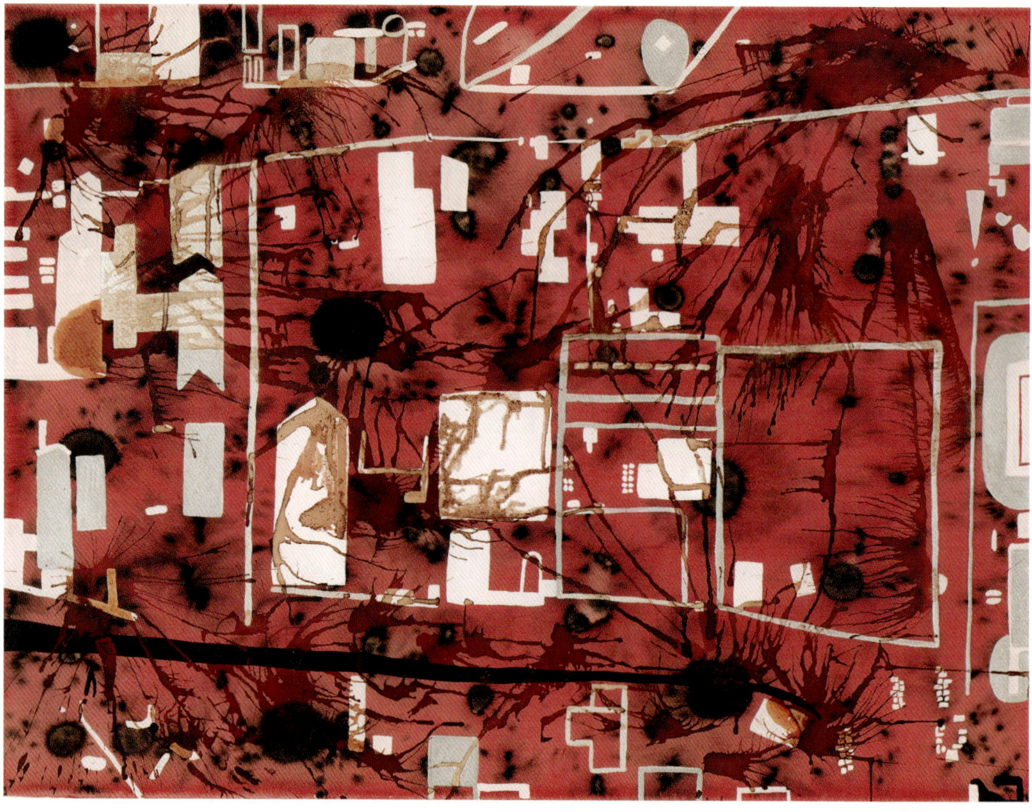

Pakistan, Democracy Now, 1998

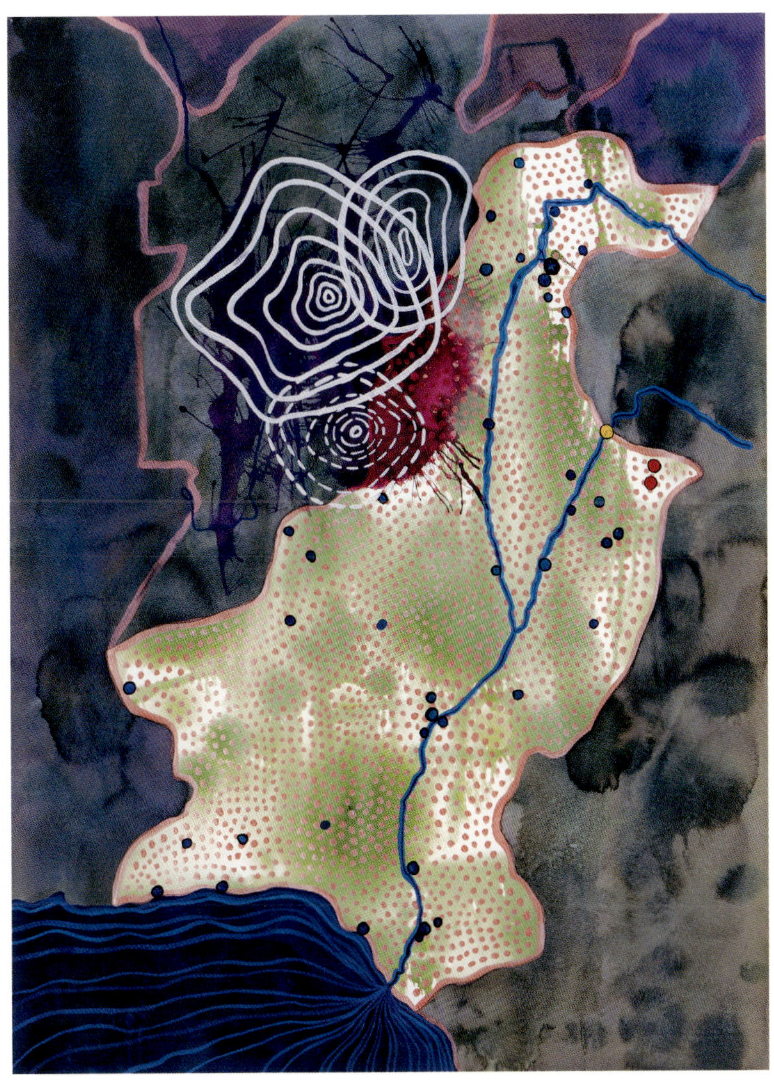

Kosovo, Yugoslavia, Operation Allied Force, 1999

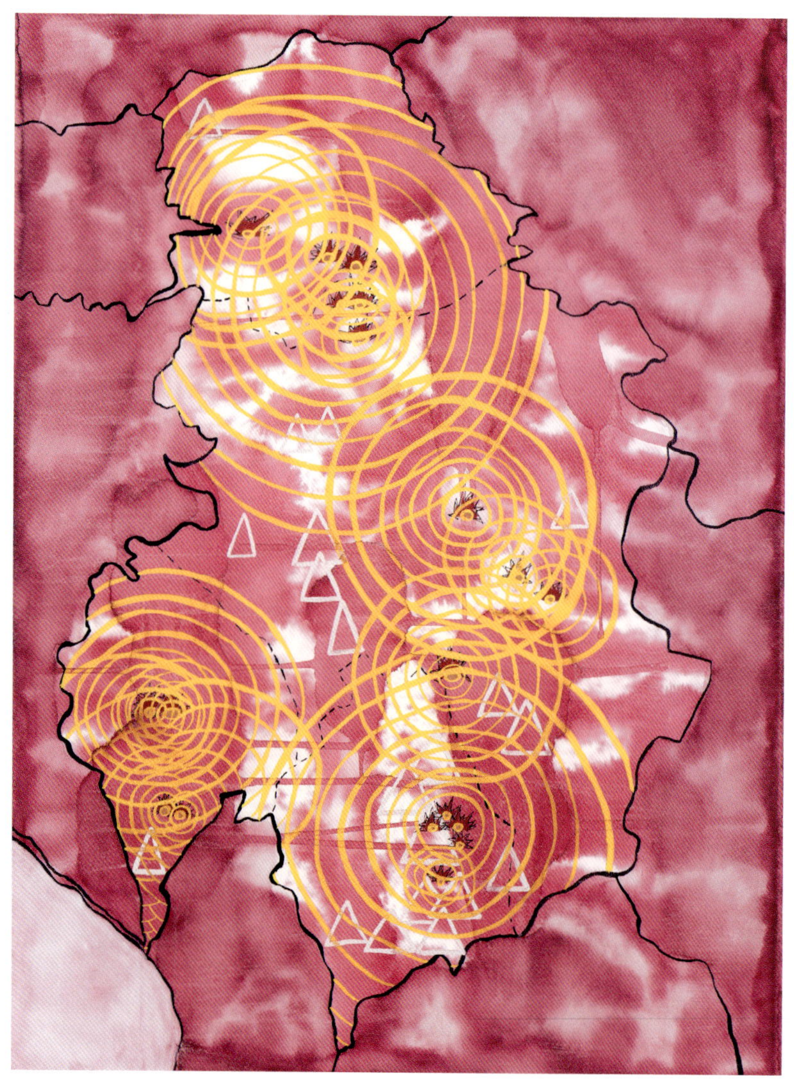

Afghanistan II, Operation Enduring Freedom, 2001–Ongoing

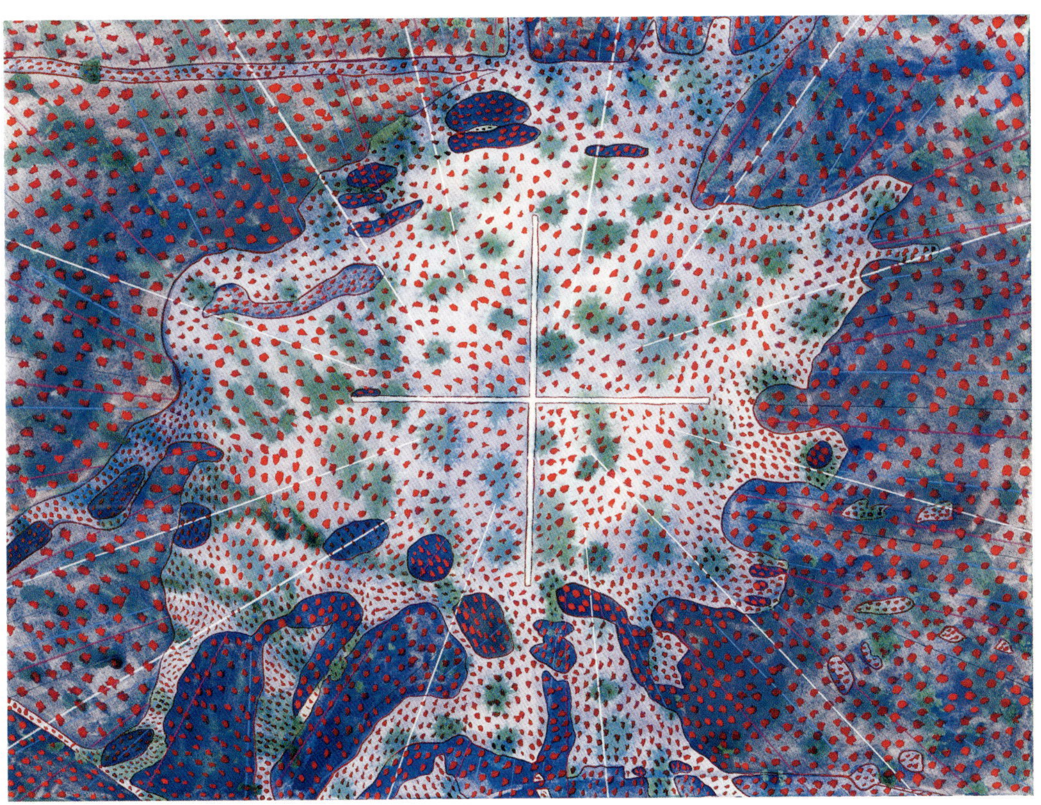

Annotated List of Plates

All drawings are mixed media on Arches papers,
30 x 22 inches, unframed

World Map, Protesting Cartography: Places the United States Has Bombed, 1854–Ongoing

Flag pins mark bombsites for which there are corresponding drawings.
p. 35

Nicaragua, 1854, 1928, and 1981–1990
For the Sandinistas.

In 1854, the American navy bombarded and destroyed the undefended city of San Juan del Norte in Nicaragua, and in 1928, the U.S. Air Force bombed guerrilla strongholds of Augusto Cesar Sandino.

"In January, 1981, Ronald Reagan took office under a Republican platform which asserted that it 'deplores the Marxist Sandinista takeover of Nicaragua.' The President moved quickly to cut off virtually all forms of assistance to the Sandinistas, the opening salvos of his war against their revolution. The contras had their own various motivations for wanting to topple the Sandinista government. They did not need to be instigated by the United States. Then the American big guns began to arrive in 1982, along with the air power, the landing strips, the docks, the radar stations, the communications centers, built under the cover of repeated joint U.S.–Honduran military exercises, while thousands of contras were training in Florida and California.
American pilots were flying diverse kinds of combat missions against Nicaraguan troops and carrying supplies to contras inside Nicaraguan territory. Several were shot down and killed. Some flew in civilian clothes, after having been told that they would be disavowed by the Pentagon if captured. Some contras told American congressmen that they were ordered to claim responsibility for a bombing raid organized by the CIA and flown by Agency mercenaries. Honduran troops as well were trained by the U.S. for bloody hit-and-run operations into Nicaragua . . . and so it went . . . as in El Salvador, the full extent of American involvement in the fighting will never be known. The contras' brutality earned them a wide notoriety. They regularly destroyed health centers, schools, agricultural cooperatives, and community centers—symbols of the Sandinistas' social programs in rural areas."
William Blum, *Killing Hope: U.S. Military and CIA Interventions since World War II*
p. 36

Firebombing of Dresden, Germany, 1939–1945

"Allied forces bombed Dresden regularly between 1939 and 1945 . . . In one day and one night of bombing, by American and British planes, the tremendous heat generated by the bombs created a vacuum, and an enormous firestorm swept the city, which was full of refugees at the time, increasing the population to a million. More than 100,000 people died."
Howard Zinn, *A People's History of the United States*
p. 37

Vieques Island, Puerto Rico, U.S. , 1941–1999

The United States has been dropping bombs on the island of Vieques for five decades, including over 267 depleted uranium-tipped bombs. As a result, Vieques is contaminated with radioactivity that contributes to a cancer rate that is twice the national average. On April 19, two errant 500-lb. U.S. bombs killed David Sanes Rodriguez, a 35-year-old civilian security guard and seriously injured four others.
p. 38

Dugway Proving Ground, Utah, U.S., 1942–Ongoing

Dugway is a "massive firing range that for fifty years was the U.S. Army testing ground for some of the most lethal chemical, biological, and nuclear weapons ever

made. A slope of mountains to the east is pockmarked with hundreds of fortified bunkers storing enough toxins to eradicate mankind. Ground water is fouled with carcinogens. This was where the cold war was waged, not in battlefields in foreign lands, but in factories, laboratories, and testing ranges."
Tony Freemantle, "The Price of Peace: The World Struggles to Purge the Poison after a Cold War Chemical Weapons Binge," *The Houston Chronicle*
p. 39

Solomon Islands, Operation Cartwheel, 1942–1945

The Solomon Islands chain consists of 992 islands and atolls to the northeast of Australia. The Solomon Islands was the scene of some of the bloodiest land, sea, and air battles of World War II. The island of Guadalcanal saw some of the fiercest fighting in the Pacific theater as the United States battled for control of the territory from Japanese occupiers through a series of Allied offensives called Operation Cartwheel.
p. 40

Nauru Island in the Gilberts, now the Republic of Nauru, 1943

The Japanese-built airfield on Nauru was bombed in 1943, preventing food supplies from reaching the island. This drawing takes as its reference an image of a B-24 Liberator bomber strike against facilities on Nauru. By late 1943, the 7th Air Force was regularly sent to "soften up" those islands scheduled for amphibious assault or to neutralize enemy forces on those islands of little strategic value.
p. 41

Poland, 1943–1944

Allied aircraft flew over 40 bombing missions in Poland. Over 900 500-lb. bombs were dropped during one mission alone.
p. 42

Assault on Iwo Jima, Japan, 1944–1945

"B-29 Superforts and B-24 Liberators pummeled the island mercilessly. Iwo Jima was bombed for 72 consecutive days, setting the record as the most heavily bombed target and the longest sustained bombardment in the Pacific War. Every square inch of the island was bombed. The 7th Air Force dropped 5,800 tons in 2,700 sorties. In one square mile of Iwo Jima, a photograph showed 5,000 bomb craters—all this on an island 5.5 miles long and 2 miles wide."
James Bradley, *Flags of our Fathers*
p. 43

Kwajalein Atoll, Republic of the Marshall Islands 1944–Ongoing

The Kwajalein Atoll was bombed during World War II in offensives against the Japanese. This drawing takes as its reference a 2001 photograph of a "Peacekeeper" missile reentry at Kwajalein Missile Range. The mission of the Ronald Reagan Ballistic Missile Defense Test Site on Kwajalein Atoll today is to provide support for Operational and Developmental Testing of Theater Ballistic Missiles, Strategic Ballistic Missiles, and Theater and Strategic Missile Interceptors.
p. 44

D-Day or Invasion Beaches, Normandy, France, Operation Overlord, 1944

Operation Overlord on June 6 was the largest military invasion of the world at the time. Allied troops sent 12,000 aircraft, 150,000 troops, 20,000 airborne troops, 5,300 ships, and 1,500 tanks. Over 425,000 Allied and German troops were killed, wounded, or went missing during the Battle of Normandy. Between 15,000 and 20,000 French civilians were killed, mainly as a result of Allied bombing.
p. 45

France, 1944–1945

"A combined air-ground attack completed the destruction of the French seaside resort of Royan, a town of ancient châteaux and lovely beaches (a favorite spot of Picasso), on the Atlantic coast near Bordeaux. A front-page dispatch in the *New York Times* from Paris reported 'two days of shattering aerial bombardment and savage ground attacks. More than 1,300 Flying Fortresses and Liberators of the U.S. 8th Air Force prepared the way for today's successful assault by drenching the enemy's positions with about 460,000 gallons of liquid fire that bathed in flames the German positions . . .' The liquid fire was napalm, used for the first time in warfare. The following day, there was another bombing, with high explosive bombs, and further ground assaults. There is no accurate count on the civilian dead resulting from those attacks, but the *New York Times* dispatch by a correspondent in the area reported: French troops mopped up most of Royan, a town of 20,000. About 350 civilians, dazed or bruised by two terrific air bombings in 48 hours, crawled from the ruins and said the air attacks had been 'such hell as we never believed possible.'"
Howard Zinn, *The Zinn Reader: Writings on Disobedience and Democracy*
p. 46

We Are Our Own Enemy, Alamogordo, New Mexico, U.S., 1945

This drawing takes as its reference an aerial photograph from the Los Alamos National Laboratory Archive of the crater formed by the first atomic explosion. On July 16, the Trinity test was the first test of a nuclear weapon conducted by the United States on what is now White Sands Missile Range. The detonation is credited as the beginning of the Atomic Age.
p. 47

Hypocenter in Hiroshima, Japan, 1945

The atomic bomb called *Little Boy* "dropped on Hiroshima on August 6, 1945, turned into powder and ash, in a few moments, the flesh and bones of 140,000 men, women, and children. Three days later, a second atomic bomb dropped on Nagasaki killed perhaps 70,000 instantly. In the next five years, another 130,000 inhabitants of those two cities died of radiation poisoning. Those figures do not include countless other people who were left alive, but maimed, poisoned, disfigured, blinded. A Japanese schoolgirl recalled years later that it was a beautiful morning. She saw a B-29 fly by, then a flash. She put her hands up and 'my hands went right through my face.' She saw a 'man without feet, walking on his ankles.'" Howard Zinn, *Hiroshima: Breaking the Silence*
p. 48

Nagasaki City at Bombing Time, Japan, 1945

On the morning of August 9, 1945, the U.S. B-29 Superfortress dropped the nuclear bomb code-named *Fat Man* over the city. The explosion generated heat estimated at 7,000 degrees Fahrenheit and winds that were estimated at 624 miles per hour. About 70,000 of Nagasaki's 240,000 residents were killed instantly, and up to 60,000 were injured.
p. 49

Boken Island, Republic of the Marshall Islands, 1945–2001

"A 1978 photo shows the crater from an atomic test at Boken Island. Americans still shape the destiny of places like Kwajalein Atoll. Several times a year, these tiny Pacific islands witness a light show more spectacular than almost any offered by nature, when a gaggle of ballistic missiles, minus their warheads, streak in low, roaring through the night sky and hurtle into the nearby lagoon . . . Thirty-three years after the last of the 66 atomic and hydrogen weapons were exploded in the area, more than 350 displaced Bikini natives are still living on Kwajalein Atoll."
Howard French, "Dark Side of U.S. Quest for Security: Squalor on an Atoll," *New York Times*, June 11, 2001
pp. 50, 51

Vienna, Austria, 1945

This drawing takes as its reference an aerial photograph of the section of Vienna where the 376th Heavy Bombardment Group used 40 aircraft, loaded with 38 tons of fragmentation bombs and 35 tons of 100-lb. bombs, to cover the entire area completely on March 22. Vienna was bombed 52 times during World War II. The raids were entirely non-military sites: the Vienna State Opera, Saint Stephen's Cathedral, the zoo, civilian neighborhoods. In February and March, 80,000 tons of bombs were dropped by U.S. and British aircraft, killing about 30,000 people and destroying more than 12,000 buildings.
p. 52

The Firebombing of Tokyo, Japan, 1945

On March 9 and 10, American B-29 bombers attack Tokyo, a city of six million people. Six hundred bombers drop 1,665 tons of napalm-filled firebombs, destroying 16 square miles. "In one horrific night, the firebombing of Tokyo—then a city largely of wooden buildings—killed an estimated 100,000 people."
Howard French, "100,000 People Perished, But Who Remembers?," *New York Times*, March 14, 2002
p. 53

Bikini Atoll, Republic of the Marshall Islands, 1946–1958

The Bikini Atoll was the location for 23 atmospheric bomb tests. The Bravo test in 1954 was the most powerful bomb ever detonated by the United States. A 15-megaton H-bomb, it vaporized three islands and spread radioactive debris over nearly 50,000 square miles.
p. 54

Enewetak Atoll, Republic of the Marshall Islands, 1948–1958

In total, there were 43 nuclear tests performed at the Enewetak Atoll. Three atomic bombs were detonated as part of Operation Sandstone in 1948. Four atomic bombs were detonated in 1951 as part of Operation Greenhouse. Two atomic tests were conducted in 1952 as part of Operation Ivy. One of them, *Ivy King*, was the largest pure-fission bomb ever detonated, with a yield of 500 kilotons, and the other, *Ivy Mike*, was the first hydrogen bomb device (it was too large to be an actual weapon), with a yield of 10.6 megatons. Six very large nuclear tests were conducted at the Bikini and Enewetak Atolls as part of Operation Castle in

1954. Seventeen nuclear weapons were detonated on the Bikini and Enewetak Atolls as part of Operation Redwing in 1956. Thirty-five weapons were detonated at the Bikini Atoll, Enewetak Atoll, and Johnston Island as part of Operation Hardtack in 1958.
p. 55

Korea, 1950–1953

"American airpower in Korea was fearsome to behold. As was the case in Vietnam, its use was celebrated in the wholesale dropping of napalm, destruction of villages, bombing cities so as to leave no useful facilities standing, demolishing dams and dikes to cripple the irrigation system, wiping out rice crops, saturation bombing, and a scorched earth policy. It is worse than useless to destroy to liberate."
William Blum, *Killing Hope: U.S. Military and CIA Interventions since World War II*

"Three years after the beginning of the war, a cease-fire was finally signed. Everything was back to where it had been in the beginning, with almost the same borders as before the war and the same unfulfilled dream of reunification. No one had won. Everyone had lost. The war is calculated to have cost the lives of 5,000,000 people, by far the majority of them civilians."
Sven Lindqvist, *A History of Bombing*
p. 56

Proving Ground, Nevada I, U.S. 1951–1992

"The Nevada Test Site was established in 1951 so the Atomic Energy Commission and the military could test nuclear weapons more cheaply and rapidly. The location was selected even though prevailing winds blowing west to east meant that tests would carry radioactivity over almost the whole United States. Furthermore, the Western Shoshone nation states that it has never ceded this land to the federal government."
Arju Makhijani, Howard Hu, and Katherine Yih, *Nuclear Wastelands: A Global Guide to Nuclear Weapons Production and its Health and Environmental Effects*

"My mother, my grandmothers, and six aunts have all had mastectomies. Seven are dead. The two who survived have just completed rounds of chemotherapy and radiation. This is my family history. Most statistics tell us breast cancer is genetic, hereditary, with rising percentages attached to fatty diets, childlessness, or becoming pregnant after 30. What they don't say is living in Utah may be the greatest hazard of all."
Terry Tempest Williams, *Refuge: An Unnatural History of Family and Place*
p. 57

Guatemala, 1953–1980

For Jennifer Harbury, wife of Guatemalan Guerilla leader Efrain Bamaca Velasquez who was captured, disappeared, and executed after torture, on order by the CIA, and for activist priest Father Thomas Melville.
"In Guatemala, in 1954, a legally elected government was overthrown by an invasion force of mercenaries trained by the CIA at military bases in Honduras and Nicaragua and supported by four American fighter planes flown by American pilots."
Howard Zinn, *A People's History of the United States*

There were many U.S. and CIA interventions between 1953 and 1980, contributing to the over 200,000 Guatemalans dead. In 1999, President Clinton acknowledged the U.S. role in the Guatemalan genocide.
p. 58

Iran, 1953 and 1987–1988

"In Iran, in 1953, the CIA succeeded in overthrowing a government which nationalized the oil industry."
Howard Zinn, *A People's History of the United States*

"Between 1987 and 1988 the U.S. Navy bombs Iran, intervening on the side of Iraq in a war."
Zolton Grossman, "A Century of U.S. Military Interventions: From Wounded Knee to Afghanistan," zmag.org
p. 59

Nevada Test Site II, U.S. 1955–1968

The drawing takes as its reference a photograph from Lawrence Livermore National Laboratory of a Project Schooner (part of Project Plowshare) crater and football field (for scale), which was made by a 35-kiloton nuclear bomb.

"From 1957 to 1973, scientists and engineers working under the direction of the U.S. Atomic Energy Commission and its Livermore laboratory investigated, experimented with, and promoted the idea of 'peaceful' uses of nuclear explosives through a program called Project Plowshare. Proponents confidently argue that the 'geographical engineering' of harbors, canals, dams, and mountain passes could be accomplished safely, economically, and scientifically by means of nuclear-blasted excavation. But the nuclear earthmoving explosions produced large amounts of radioactive fallout, and with it, significant challenges to the program on scientific grounds. The Project Schooner cratering explosion produced the highest levels of fallout in Utah recorded since 1962, and sent radioactive debris across the Canadian border."

Scott Kirsch, "Peaceful Nuclear Explosions and the Geography of Scientific Authority," *Professional Geographer*
pp. 60, 61

Johnston Atoll, U.S. 1958–1962

Thirty-five weapons were detonated at the Bikini Atoll, Enewetak Atoll, and Johnston Atoll as part of Operation Hardtack in 1958. Thirty-six atomic weapons were detonated at Johnston and Christmas (Kiritimati) Islands in 1962 as part of Operation Dominic.
p. 62

Haiti, 1959

For Jean-Bertrand Aristide, Stan Goff, and my father.
"The U.S. military mission in Haiti, to train the troops of noted dictator Francois Duvalier, used its air, sea, and ground power to smash an attempt to overthrow Duvalier by a small group of Haitians aided by some Cubans and other Latin Americans."
William Blum, *Rogue State: A Guide to the World's only Superpower*
p. 63

Peru, 1960–1965

For Hugo Blanco.
"CIA official Victor Marchetti said that Green Berets participated in what was the CIA's single large-scale Latin American intervention of the post-Bay of Pigs era ... The American objective in Peru—to crush a movement aimed at genuine land reform and social and political changes—was identical to its objectives in Vietnam. The methods employed were similar: burning down peasants' huts and villages to punish support for the guerillas, defoliating the countryside to eliminate guerrilla sanctuaries, saturation bombing with napalm and high explosives, even throwing prisoners out of helicopters."
William Blum, *Killing Hope: U.S. Military and CIA Interventions since World War II*
p. 64

The Belgian Congo, Zaire, now the Democratic Republic of the Congo, 1960–1964

For Patrice Lumumba and Barbara Kingsolver.
"U.S. Air Force C-130s were flying Congolese troops and supplies against the Katangese rebels, while the CIA was simultaneously aiding the very same rebels with aircraft and mercenary units. Washington dispatched 100–200 military personnel to the Congo, and the CIA was conducting paramilitary campaigns against the insurgents. CIA pilots carried out regular bombing and strafing missions."
William Blum, *Killing Hope: U.S. Military and CIA Interventions since World War II*

"The United Sates was covertly involved in suppressing a rebellion in the Katanga province of the Congo. At one point there were Congolese refugees streaming into western Tanzania where I was teaching. They claimed their villages had been napalmed by unmarked U.S. planes."
Randy Kehler, draft resister, cofounder of the Nuclear Weapons Freeze Campaign, from Christian Appy, *Patriots: The Vietnam War Remembered from All Sides*
p. 65

Vietnam, 1961–1971

"North Vietnam had been the target of secret U.S. military operations since 1961. In early 1965, just months after his landslide victory over Goldwater, Lyndon Johnson launched Operation Rolling Thunder, the sustained bombing of North Vietnam. In June, huge American B-52 stratofortresses began bombing targets in South Vietnam, each plane dropping up to 27 tons of explosives per mission . . . Of the 53,193 American military fatalities in Vietnam, more than 10,000 were classified as 'noncombat deaths.' U.S forces also suffered more than 300,000 wounded. In 1995, the Vietnamese government announced that 1.1 million Communist troops were killed and 600,000 were wounded during the American War. The hardest figures to find reliable estimates for are the number of Vietnamese civilians killed during the war. Most estimates indicate that civilian deaths exceed combat deaths. The Vietnamese government estimates that two million Vietnamese civilians died during the American War. In total, the war extinguished some three million lives."
Christian Appy, *Patriots: The Vietnam War Remembered from All Sides*
pp. 66, 67

Kiritimati or Christmas Island, Republic of Kiribati, 1962

Twenty-four U.S. atomic bomb tests were conducted on Kiritimati as part of the Dominic Series.
p. 68

Laos, 1964–1973

"Between 1965 and 1973 more than two million tons of bombs rained down upon the people of Laos, more than the U.S. had dropped on both Germany and Japan in WWII."
William Blum, *Killing Hope: U.S. Military and CIA Interventions since World War II*

"Fred Branfman was one of the first Westerners to expose the fact that the United States had been secretly bombing northern Laos since 1964. Many of the attacks focused on

the Plain of Jars, a high plateau controlled by the Laotian Communists, the Pathet Lao. It had been populated by some 50,000 Laotian peasants who lived among six-foot-tall vessels believed to be ancient funeral urns—the famous jars of the plain. Branfman writes, 'Every single one of the refugees told the same basic story. The bombing started in May 1964, slowly accelerated, and got really bad in 1968. I later found out that when Johnson declared the bombing halt over North Vietnam just prior to the November 1968 election, they simply diverted all the planes into northern Laos. Then Nixon and Kissinger came in and leveled the entire Plain of Jars.
Many of the air strikes dropped 'pineapple bombs'—antipersonnel bombs. They couldn't destroy trucks or antiaircraft emplacements. They were only meant to kill people. They'd shoot 250 pellets across an area the size of a football field. Then they refined them with these little fleshettes so they'd enter your body and were almost impossible to remove.'"
Christian Appy, *Patriots: The Vietnam War Remembered from All Sides*
p. 69

Mississippi, U.S. 1964–1970

The Salmon and Sterling Nuclear Test Sites were the locations for two nuclear detonations performed in a subterranean salt dome formation in Mississippi, as part of an Atomic Energy Commission Test. The test program, Project Dribble, called for creating an underground cavity, using a nuclear bomb to do so, then later detonating a second nuclear device, as well as two gas explosives, inside the cavity. The first detonation, to form the cavity, code-named Salmon, took place in 1964 using a 5.3-kiloton bomb. The second nuclear blast, a relatively small 0.38 kilotons yield shot code-named Sterling, was exploded within Salmon's 110-foot diameter cavity more than two years later.
p. 70

Amchitka Island, Alaska, U.S. 1965–1971

"This midway point on the great arc of Alaska's Aleutian Islands was designated a national wildlife refuge in 1913. The Pentagon and the Atomic Energy Commission went looking for a place to blow up H-bombs. As a result, Amchitka was the site of three large underground nuclear tests, including the most powerful nuclear explosion ever detonated by the United States. In 1965, the Long Shot test exploded an 80-kiloton bomb. In 1969, the Milrow Nuclear test was of one megaton, ten times more powerful than Long Shot. In 1971, the 5-megaton Cannikan Bomb was exploded, 385 times as powerful as the bomb dropped on Hiroshima.
Independent research by Greenpeace and newly released documents from the Department of Energy show that the Amchitka tests began to leak almost immediately. Highly radioactive elements and gasses, such as tritium, americium-241, and plutonium, poured out of the collapsed test shafts, leached into the groundwater, and worked their way into ponds, creeks, and the Bering Sea."
Jeffrey St. Clair, *In These Times*
p. 71

Cambodia, 1969–1973

"From 1962 through 1973 the United States struck Vietnam, Laos, and Cambodia with eight million tons of bombs, more than three times the amount it dropped in World War II. U.S. Policy in Cambodia from 1969 to 1973 played a crucial, if unintended, role in building support for the Cambodian Communists (the Khmer Rouge). A once insignificant rebel group, the Khmer Rouge did not gain strength until the United States began secretly bombing Cambodia."
Christian Appy, *Patriots: The Vietnam War Remembered from All Sides*
pp. 72–73

Afghanistan I, 1979 and Infinite Reach, 1998

"The striking repression of women in Afghanistan carried out by the Taliban Islamic Fundamentalists is well known. Much less publicized is that in the late seventies and most of the eighties Afghanistan had a government committed to bringing the incredibly underdeveloped country into the twentieth century, including giving women equal rights. The U.S., however, poured billions of dollars into waging a terrible war against this government, simply because it was supported by the Soviet Union. By aiding the fundamentalist opposition, Washington knowingly and deliberately increased the probability of a Soviet intervention. And when that occurred, the CIA became the grand orchestrator. In the end, the U.S. and the Taliban 'won,' and the women and the rest of Afghanistan, lost: more than a million dead, three million disabled, five million refugees, in total about half the population."
William Blum, *Rogue State: A Guide to the World's only Superpower*

In August 1998, the United States announced Infinite Reach, anti-terrorist missile strikes against targets in Afghanistan and the Sudan. Seventy-five to 100 missiles were launched, leaving 21 dead and 50 wounded.
p. 74

El Salvador, 1980–1994

For Archbishop Oscar Romero.
"A campaigner against the Salvadorean army's death

squad war, Monsignor Romero was shot through the heart while saying Mass, shortly after appealing to the U.S. not to send military aid to El Salvador. The appeal fell on deaf ears and for the next 12 years, the U.S. became involved in its largest counter-insurgency war against left-wing guerrillas since Vietnam. To defeat the rebels, the U.S. equipped and trained an army which kidnapped and disappeared more than 30,000 people, and carried out large-scale massacres of thousands of old people, women, and children."
BBC News, March 24, 2002, Tom Gibb, BBC correspondent in El Salvador during the eighties
p. 75

Lebanon, 1983–1984 and 2006

For Rania Masri.
". . . the U.S. bombardment of Beirut in 1983 and 1984."
William Blum, *Rogue State: A Guide to the World's only Superpower*

"President Bush and Vice-President Cheney were convinced, current and former intelligence and diplomatic officials told me, that a successful Israeli Air Force bombing retaliatory attacks campaign against Hezbollah's heavily fortified underground-missile and command-and-control complexes in Lebanon could ease Israel's security concerns and also serve as a prelude to a potential American preemptive attack to destroy Iran's nuclear installations, some of which are also buried deep underground. 'The Israelis told us it would be a cheap war with many benefits,' a U.S. government consultant with close ties to Israel said. 'Why oppose it? We'll be able to hunt down and bomb missiles, tunnels, and bunkers from the air. It would be a demo for Iran.'"
Seymour Hersch, "Watching Lebanon: Washington's Interest in Israel's War," *The New Yorker*, August 21, 2006

"The U.S. State Department writes, in its evacuation letter to its citizens, 'The U.S. Embassy reaffirms the firm, enduring, and nonnegotiable commitment of the United States to Lebanon and the Lebanese people.' Simultaneously, the U.S. speeds up bomb delivery for the Israelis, as reported in the *New York Times*. Simultaneously, Israel declares that in the past 24 hours Israel has hit more than 140 'targets' in Lebanon—with U.S. weaponry. It is clear. This isn't simply Israel's war on Lebanon, the country, and the people. This is the U.S./Israel's war on Lebanon, the country, and the people. I would rather the U.S. not have an 'enduring commitment' to Lebanon. More than 1,000 Lebanese have been killed, 30% of them children. Four thousand have been wounded and one million have been displaced. Thirty thousand housing units and over 70 bridges and 94 roads have been destroyed and 100,000 housing units have been rendered uninhabitable."
Rania Masri, unpublished speech, 2007
p. 76

Grenada, Operation Urgent Fury, 1983

The Invasion of Grenada left over 100 American, Grenadian, and Cubans dead. "Abuse took on unsuspected proportions in the Caribbean when the U.S. bore down on a tiny country which is less than one twenty-seven-thousandth its size, with an army much smaller than the police force of any U.S. city, with a population which is less than that of a New York City borough, and on whose land mass it was difficult even to deploy all the military machinery used in the invasion."
Raul Castro, *Grenada: The World against the Crime*
p. 77

Philadelphia, the Firebombing of M.O.V.E., U.S. 1985

For Ramona Africa and Mumia Abu-Jamal.
On May 13, 11 men, women, and children are burned alive, along with an entire city block, when the Philadelphia Police Department dropped an incendiary bomb (200 lbs. of gasoline with a detonator) from a police helicopter onto the roof of M.O.V.E.'s building because the separatists refused to come out and face the police firing squad.
p. 80

Libya, Operation El Dorado Canyon, 1986

"In April, a bomb exploded in a discotheque in West Berlin, killing two people, one an American soldier. It was unquestionably an act of terrorism. Libya's tyrannical leader, Muammar Khadafi, had a record of involvement in terrorism, although in this case there seemed to be no clear evidence of who was responsible. Nevertheless, President Reagan ordered that bombers be sent over Libya's capital of Tripoli, killing perhaps a hundred people, almost all civilians."
Howard Zinn, *The Zinn Reader: Writings on Disobedience and Democracy*
pp. 78, 79

Panama, Operation Just Cause, 1989–1990

"A full-scale military operation involving 24,000 troops with high-tech weaponry created terror in the middle of the night, and for weeks after, in every city and rural area in Panama. As the report of the Commission shows, from 1,000 to 4,000 Panamanians were killed, thousands more wounded, and more than 20,000 found themselves homeless. Thousands more Panamanians were arrested, including union leaders, university professors, political organizers, government officials, journalists, and military personnel. All those

arrested had one thing in common. Each was known to be a defender of Panamanian independence and sovereignty and opposed the U.S. invasion." Independent Commission of Inquiry on the U.S. Invasion of Panama, *The U.S. Invasion of Panama: The Truth Behind Operation "Just Cause"*
p. 81

Baghdad, Iraq, 1990–Ongoing

This drawing takes as its reference a newspaper map showing the targets hit during the first 24 hours of the first "Gulf War" in 1990. ". . . dropping 177 million pounds of bombs on the people of Iraq in the most concentrated aerial onslaught in the history of the world." William Blum, *Killing Hope: U.S. Military and CIA Interventions since World War II*

"Every day for years after Gulf War I, British and U.S. warplanes bombed Iraq, and civilians and children died. Children also died every day in Iraq because of the U.S. sponsored embargo. Then George W. Bush came to the White House and invaded again, with civilians in Iraq now being killed at the rate of more than 3,000 *per month.* The loss of life among Iraqi civilians is especially startling. In November 2004, the British medical journal *The Lancet* reported that up to 100,000 civilians had died so far as a result of the war, many of them children." Howard Zinn, *A Power Governments Cannot Suppress*
p. 82

Kuwait, 1991

Referring to depleted uranium-tipped missiles, "'The United States,' wrote international environmental activist Dr. Helen Caldicott, 'has conducted two nuclear wars. The first against Japan in 1945, the second in Kuwait and Iraq in 1991.'"
William Blum, *Rogue State: A Guide to the World's only Superpower*
p. 83

Shifa Pharmaceutical Plant, Khartoum, Sudan, 1998

On August 20, the "U.S. Military Strikes on a Chemical Weapons Plant," as part of Infinite Reach, in retaliation for "terrorist attacks" on American embassies in Tanzania and Kenya. The U.S. government inaccurately thought the plant was producing chemical weapons for Osama Bin Laden. The pharmaceutical plant had raised Sudanese medicinal self-sufficiency from less than 5% to more than 50%, while producing 90% of the drugs used to treat the most deadly illnesses in this desperately poor country. A dozen Tomahawk cruise missiles instantly deprived the people of Sudan of this achievement.
p. 84

Pakistan, Democracy Now, 1998

President Clinton orders cruise missile attacks on targets in Afghanistan and The Sudan. Seventy-five to 100 missiles are launched and at least one errantly lands in Pakistan, killing five people. "So we went to war first against Afghanistan and then Iraq, and have bombed locations in Pakistan and Yemen, inevitably killing innocent people because it is in the nature of bombing—and I say this as a former U.S. Air Force bombardier—to be indiscriminate, to 'make no distinction.'"
Howard Zinn, *A Power Governments Cannot Suppress*
p. 85

Kosovo, Yugoslavia, Operation Allied Force, 1999

"After pounding about 40 targets across Yugoslavia with cruise missiles and smart bombs during the first wave of NATO attacks Wednesday night, dozens of U.S. and Allied planes, ships, and submarines launched a new barrage Thursday night. No official information on which targets were hit was available. The targets marked were reported by witnesses and Yugoslav and other media reports."
The Washington Post, Friday, March 26, 1999
p. 86

Afghanistan II, Operation Enduring Freedom, 2001–Ongoing

This drawing takes as its reference an Associated Press/Department of Defense photograph made from a gun camera video that was released during a Pentagon briefing in Washington, October 11, 2001. It shows a target as a blast erupts during a strike by U.S. forces on a surface-to-air missile site located within Afghanistan in retaliation for the September 11 attacks on the U.S. "Thousands of civilians in Afghanistan have already been killed as a result of our military operations, and countless thousands more are being wounded, losing limbs, blinded. Often these are children, victims of unexploded land mines, or cluster bombs. But the American people are not told these stories; we are kept ignorant of what the 'war on terror' means in human terms. Sayed Salahuddin of Reuters reported on October 28, 2001, from Kabul: 'A U.S. bomb flattened a flimsy mud brick home in Kabul Sunday, blowing apart seven children as they ate breakfast with their father. Sobs racked the body of a middle-aged man as he cradled the head of his baby, its dust-covered body dressed only in a blue diaper, lying beside the bodies of three other children, their colorful clothes layered with debris from their shattered homes.'" Howard Zinn, *A Power Governments Cannot Suppress*
p. 87

What We Cannot See: An Interview with elin o'Hara slavick

Catherine Lutz

Catherine Lutz: Could you describe your drawings and the process of making them, and could you speak about why you chose abstraction and the aerial view?

elin o'Hara slavick: One common formal element of the drawings is that each one begins with ink or watercolor dropped onto wet paper like bloodstains on damp clothing. When it dries this becomes the foundation upon which to tell a violent story. I use this ground of abstract swirling or bleeding to depict the manner in which bombs do not stay within their intended borders. Depleted uranium and chemical agents contaminate the soil, traveling in water and currents of air for decades. Mines and unexploded bombs lay in wait for unsuspecting victims who were not even alive during the war. Bombs lay the groundwork for genocide, cancer, more war, terrorism, widows, orphans, and a vengeful populace on all sides of conflict.

The drawings are relatively abstract—and I say relatively because there *are* some recognizable cartographic, geographic, and realistic details like arrows, borders, and airplanes—and as in war, civilians are rendered invisible. I employ abstraction to reach people who might otherwise turn away from realistic depictions. People approach abstraction with fewer expectations and defenses. I want to reach people who have not made up their minds, who long for more information, the people who vote and want to believe that we are living in a democracy, but are filled with fear and doubt.

The drawings are also beautifully aerial to seduce and trap the potentially apathetic viewer, so that she will take a closer look, slow down, and contemplate the accompanying information that may implicate her. I also chose the aerial view to align myself, as an American, with the pilots dropping the bombs, even though I would not drop them. As a photographer aware of the military's use of the aerial view that flight and photography provide, using the aerial view seems like the natural choice. I utilize surveillance imagery, military sources and battle plans, photography and maps, much of which is from an aerial perspective.

CL: Given that you have already used photography in a powerful way to comment on militarization in *Homefront: A Military City and the American 20th Century*, what made you choose a different medium for this project?

eos: Making the documentary photographs for *Homefront* was done to illustrate your text. Some of the photographs are strong enough to stand on their own, but most of them are only meaningful alongside your words. The bomb drawings rely on the lengthy titles as well to convey their meaning. However, each one was finished as a painting first and *then* titled.

I chose drawing because of my ongoing struggle with the problematic nature of photography. While the drawings are not photographs, they *are* photographic. Many of them are drawn from photographic sources and most of them are from an aerial perspective that is inherently photographic. But I cannot make photographs of these damaged places. I did not survive the bombings as a victim, but as a war-tax-paying citizen of the bombing nation. Even if I *could* make photographs, I would not because there are already too many photographs—too immediate, too true, too real, and too brief—countries, and lives reduced to singular images. I hope that if I labor on a series of drawings in which the artist's hand is visible, that people will work to understand them on a deeper and more complicated level than they might when seeing a photograph. Inspired and informed by documentary photographs and violent maps, I want to convey the unsignifiable, to offer a protest against meaninglessness, to do *something* in the face of so much destruction. John Berger's ceaseless calls for hope and art challenge me while being fully aware of the impossible reconciliation between art and the reality of such horrific events that W.G. Sebald and Kyo Maclear write so eloquently about. Making these drawings is just one aspect of my life as an activist. I have a similar hope when I organize and march against war—that people will reconsider the present through history to affect the future.

However, I *do* worry about the use of abstraction to address such a magnitude of destruction. W.G. Sebald's *On the Natural History of Destruction*—a book about the ceaseless Allied bombing of German cities and the subsequent little space it occupies in Germany's cultural memory and the failure of the German people to represent it—both challenges and paralyzes me. Sebald writes: '. . . the construction of aesthetic or pseudo-aesthetic effects from the ruins of an annihilated world is a process depriving literature of its right to exist.' What then is an artist to do? Should I put these drawings away? Should I display images of shriveled and burned corpses, photographs of the guilty military generals, pictures of the ruins next to the drawings? I am troubled by these very serious questions, but I think I have reached many people who may have otherwise walked away from realistic depictions of war. As Sebald also writes: 'The issue, then, is not to resolve, but to reveal the conflict.'

CL: Perhaps you are responding, too, to what Paul Virilio has identified as the intensifying speed of warfare during the twentieth century. If bombing, like other war technologies, is a primary engine of history, as he claims, and if 'history progresses at the speed of its weapons systems,' how can your art or how has other art responded to this problem?

eos: I feel defeated by this idea even though I think it's true. I am more interested in Virilio's claim that the 'true enemy' becomes 'less external than internal: our own weaponry, our own scientific might which in fact might promote the end of our own society.' We are racing against ourselves, bombing ourselves. I am more afraid of my own government than I am of 'terrorists.' And as John Armitage writes in the introduction to Virilio's *Art and Fear*: 'When Virilio considers the aesthetics of disappearance, he assumes that the responsibility of artists is to recover rather than discard the material that is absent and bring to light those secret codes that hide from view inside the silent circuits of digital and genetic technologies.' Indeed, Virilio's book, *Art and Fear*, is a chilling book for an artist to read. Like Sebald, Virilio makes me feel as if there is no manner in which to properly, effectively, or truthfully represent or respond to the horrors of war, genocide, and the holocaust. Yet, they both attempt to do just that and they do it very well.

Drawings function differently than words in a book and the endless stream of information on the web. Not necessarily narrative or full of facts and figures, drawings are a *visual* interpretation or depiction of, reaction against, reflection on, and emotional response to the world around us. While some artists—mostly those working with electronic media—attempt to keep up with the 'intensifying speed' of everything around us—warfare, information, and communication systems—many artists, like me, are responding by trying to slow down, not only ourselves, but viewers as well. I want to slow people down and offer a different approach to seeing and thinking about war. As Kyo Maclear writes in her book, *Beclouded Visions: Hiroshima-Nagasaki and the Art of Witness*: 'When one considers how our contemporary image culture is evolving amid real-time suffering, the question of witnessing takes on a renewed importance. Our death-saturated world continues to produce numerous collective traumas, traumas demanding historical awareness, yet often defying our usual modes of access. These collective traumas suggest the need for a prolonged gaze. The televisual gaze, however intense, is all too brief.'

Specifically, I have responded by completing over forty-five of these drawings, and when they are all exhibited together, it is a dizzying experience: history comes all mixed up together, Korea beside Germany, Hiroshima next to Kosovo. Utter disbelief and shame that 'we' have bombed so many places and, yet, the series can never be completed. There is always another war. Alfredo Jaar employs various means to interrupt the impossible speed of warfare. In one installation, Jaar displays photographs of a war-torn country in black boxes with bare burning light bulbs dangling above them. Viewers are not allowed to open the boxes, depriving us of the all-too-familiar voyeuristic experience of horror. It is an installation of darkness, death, and mourning—one that implicates us as blind participants. Colombian artist Doris Salcedo is another artist who works slowly and who slows us down as viewers. She sews hair into tables and buries shoes belonging to 'the Disappeared' in walls. It is as if she is making work from the grave.

CL: The other change in how war comes to be known by the general public is the result of an intensifying use of massive public relations or propaganda tools by the Pentagon for recruitment and for retaining popular support for its activities. It invests, by some estimates, several billion dollars annually on this, and its images, too, are often beautiful. How do your paintings enter this difficult—some might even say impossible—arena of war making as public relations?

eos: I do not kid myself. My drawings function primarily in the art world. That said, you might be surprised by the art world's easy patriotism. I am not convinced that I am 'preaching to the converted.' In this realm, my drawings are antirecruitment posters; protests against bombing; a propaganda campaign against war; a blatant critique of U.S. foreign policy and activities. I think *all* art enters the arena of public relations or propaganda to some extent. All art wants to persuade us to enter into the world the artist presents.

While most of my drawings are done from military sources, they are very different than their sources. I delete most of the revealing information—names of cities and geographical markers—and try to disengage these places from authority's clenched fist.

As a university professor, I can attest to the power of art to educate. I believe that education may be the most powerful form of public relations, and if we can interject a discussion of peacemaking and an understanding of U.S. involvement in war after war after war into the deafening noise of patriotism and the chaotic speed of information and communication systems, then we have begun to challenge the status quo.

CL: Your drawings are deeply historical. Like critical historiography, they tell a very different kind of story about American military history and global history. What other more accepting fields of war art do these drawings enter into? In other words, what other art do these paintings protest?

eos: I decided to call the series *Protesting Cartography: Places the United States Has Bombed* precisely because I think they are more than just antiwar drawings. They protest the age-old power of maps; power utilized by governments and individuals in the name of private ownership, border control, and imperialism.

They are also in protest of and in reaction to the tired arguments against political art and for the purity of form, as if form and content are not ultimately married. 'Good art is not political' or 'political art is usually bad art' or 'art does not change the world' is how the arguments go. What about Goya's timeless *Disasters of War*, Picasso's sensational *Guernica*, Lewis Hine's haunting photographs of child laborers that brought about child labor laws, and Sue Coe's gripping and graphic drawings of slaughterhouses, war, AIDS, and poverty? I watch art change students' lives every semester.

In hindsight, perhaps it is exactly those conservative arguments made against and out of fear of such powerful art that drove me to choose a more abstract and painterly form. While one could argue that American abstract expressionism was covertly about big, bad, post-World War II, militaristic America, Jackson Pollock's splatter paintings are primarily still discussed as action paintings. They are all about the energy and motion of the artist, the paint on the canvas. They are in reaction to and about art itself. My drawings react against this 'art for art's sake,' while utilizing one of its forms.

CL: Your paintings bring many places to view that will be familiar to people who have learned from their school textbooks that American history is the history of war in places like Europe and Korea and Iraq. But you have as many or more paintings that depict places that have suffered their bombing in silence and invisibility—places like Enewetak Atoll, the Sudan, or Mississippi. Did you find different challenges in drawing these two kinds of places?

eos: Not formally. I tried to keep them somewhat similar visually. I am troubled by having only one drawing of Vietnam, Cambodia, and Hiroshima, just as there is only one drawing of Mississippi, Alaska, and the Sudan. Some countries were bombed relentlessly and ruthlessly for decades. Other places we tested nuclear weapons. Other places were accidents. I do not mean to equalize these bombings. Each one has its own history.

And it is important to show places relatively unknown, like Enewetak Atoll, alongside infamous sites like Iwo Jima. It is imperative that I not only draw the places familiar to everyone, but include the lesser known locations, so that people can make connections and begin to have a sense of the unimaginable scope of our violence against civilians, even against ourselves.

CL: Yes, the places the U.S. has bombed include domestic spaces as well: your maps include Mississippi, Nevada, and Amchitka, Alaska, among other domestic sites. We generally do not think of ourselves as the victim of ourselves, or see the immense costs of war preparation at home. How have people who view your exhibits responded to what is in all likelihood the shocking news that America has bombed itself?

eos: It certainly is important that I include domestic places and in some ways, these have proved to be the most persuasive. People start to question the very foundation and purpose of war, and begin to feel some solidarity with those under the bombs.

One of the domestic drawings, *Philadelphia, the Firebombing of M.O.V.E.*, was included in the traveling exhibition *Toxic Landscapes: Artists Examine the Environment.* The exhibition traveled throughout the U.S. and to the Jose Marti National Library in Havana, Cuba. Not surprisingly, more Cubans appear to know about this recent firebombing of American citizens by American police than Americans do.

I must admit to having been shocked myself when I discovered that we had bombed ourselves repeatedly and in so many different places: Mississippi, Nevada, Utah, Alaska, Puerto Rico, Philadelphia, New Mexico, Colorado, and North Carolina, to name the sites for which there are drawings. I knew about the Nevada test sites, but had no idea that we detonated huge nuclear test bombs in Alaska. As Howard Zinn and others have written, bombs are dropped because they have been made and it would be a waste of expended resources and investments *not* to test and use them.

CL: Earlier you spoke about the need for a more prolonged view of the impact of bombing: your drawings suggest both the moment of bombing or trauma itself and, as you've noted, your techniques suggest the longer term impacts of things like the toxins that remain behind. Those represent another form of bombing, at the cellular level, that have left the people of Vietnam and Guam, and Vieques, for example, with graveyards filled with victims of the resulting cancers and other diseases. Do your drawings or could other drawings look more closely at this other ongoing result of bombing?

eos: I *do* hope that the cellular references that appear in many of the drawings—replicated stains in the background, connected tissue in the foreground, concentric targets like microscopic views of damaged cells—conjure up the buried dead and deadly diseases as a result of warfare.

Some of the drawings are accompanied by titles with information about the illnesses and deaths as a result of the bombings. While the drawings do not literally contain this information, it is implied. We know that more civilians die in war than combatants. We know that uranium tipped missiles cause radiation-related diseases like cancer and that landmines remove limbs from innocent children years after the conflict.

Carole Gallagher's book *American Ground Zero: The Secret Nuclear War* uses photographic portraits of and essays on communities living within and downwind of America's test sites in the southwest to reveal the duplicity of the U.S. government in creating one of the most carcinogenic landscapes in the world. Perhaps, in this instance, photography and the written word are more effective.

CL: You have worked tirelessly to find and draw each of these places of horror, but in the end, there are many more left undrawn and, unfortunately, more such places being made everyday around the globe. Can you say more about the places you have not drawn? If you or another group of artists were to continue this project, where would you go?

eos: I keep the written list going. I still want to do one of Cuba and the Philippines. I have not made a new bomb drawing for a few years now, ever since the birth of my first child. I just had to stop. The last one I made was a map of the world with flag pins at every place for which there is a corresponding drawing. I did it for the exhibition *Violent Violence* that I organized of American artists dealing with violence in their work at Arti et Amicitiae in Amsterdam. Maybe someday I will work on a separate, but related, series on individual countries—a hundred drawings of Iraq, a thousand of Vietnam. I have a feeling I'll be working on this series for years to come.

I want to work on something utopic. What would the opposite of a ruined city look like? Where is that? In Greifswald, Germany, as I write this, I am struck by the camouflage of history. I want to photograph this odd transitional place—a transition from East Germany to Germany, from the East to the West, from socialism to capitalism, from the past to the present, from failure and defeat to reconstruction and resignation. Old drab buildings are boarded up, torn down, renovated, or painted pale orange. How can people be riding their bikes over land that once was drenched with civilian blood as if nothing happened? The erasure of history and the constant denial of the ongoing trauma beg for representation. I will continue to try to make visible the traces of war and peace, history and resistance to it, the state of the world as we know it and destroy it.

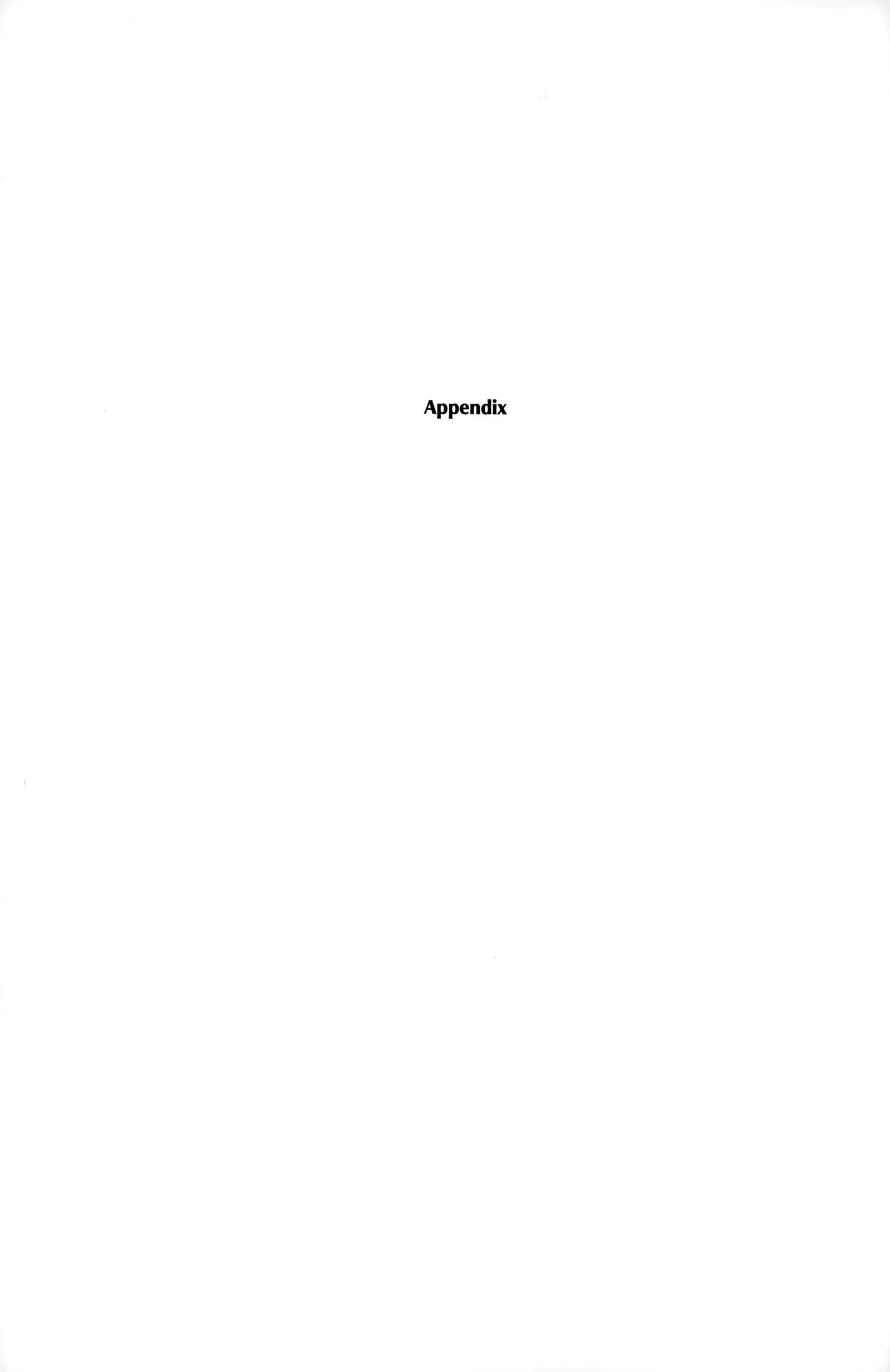

Appendix

Biographical Notes

About the Artist

Elin o'Hara slavick was born in 1965 in New York and grew up in Portland, Maine. The youngest of seven children, she was raised in a radical Catholic activist environment. She studied at Sarah Lawrence College and The School of the Art Institute of Chicago, where she received her MFA in Photography in 1992. She is an artist and Professor of Studio Art, Theory, and Practice at the University of North Carolina at Chapel Hill. She has exhibited her work across the United States and in Canada, Europe, Cuba, Hong Kong, England, and Scotland. She lives in Chapel Hill with her husband, epidemiologist David Richardson, and their two children.
www.unc.edu/~eoslavic

About the Contributors

Catherine Lutz is a Professor of Anthropology at Brown University. She is the author of *Homefront: A Military City and the American 20th Century* (with photographs by elin o'Hara slavick); *Unnatural Emotions*; and coauthor with Jane Collins of *Reading National Geographic*.

Carol Mavor is currently Professor of Art History and Visual History at the University of Manchester, UK. She is the author of three books: *Reading Boyishly: Roland Barthes, J. M. Barrie, Jacques Henri Lartigue, Marcel Proust, and D. W. Winnicott*; *Becoming: The Photographs of Clementina, Viscountess Hawarden*; *Pleasures Taken: Performances of Sexuality and Loss in Victorian Photographs*, all published by Duke University Press. Currently, she is finishing a novel entitled *Full*, and a slim film book entitled *Black and Blue*.

Howard Zinn is a decorated World War II bombardier. In 1968, he flew to Hanoi with Father Daniel Berrigan to receive the first three American fliers released by North Vietnam. An activist, playwright, and historian, he has written over two dozen books, including: *The Politics of History*; *Vietnam: The Logic of Withdrawal*; *Just War*; *A Power Governments Cannot Suppress*; and his epic masterpiece *The People's History of the United States*. Zinn is a Professor Emeritus of Political Science at Boston University. He lives with his wife Roslyn in the Boston area, near their children and grandchildren.

Exhibitions

Selected Solo Exhibitions

2002
Workers Dreaming and Protesting Cartography, The Annex, New York.
Protesting Cartography, John Hope Franklin Center, Duke University, Durham, North Carolina.

2000
New Frontiers IV, The Mint Museum, Charlotte, North Carolina.

1999
Workers Dreaming, with Jane Marsching, Brewhouse Space 101, Pittsburgh.
Bragg Murders - A Violent Cartography, Modern Museum, Durham, North Carolina.

1997
Labour < > Leisure (Global Economy), Lump Gallery, Raleigh, North Carolina.

1996
mutter, The Other Gallery, Banff Centre for the Arts, Alberta, Canada.

1992
A Wall of Incoherent Dresses, Franklin Furnace, New York.

Selected Group Exhibitions

2007
The Big Picture, Contemporary Photography, North Carolina Museum of Art, Raleigh, North Carolina.

2006
New Cartography, LoBot Gallery, Oakland, California.
Restating Empire, Gallery 110 C, Seattle.
Crosscurrents, Mint Museum of Art, Charlotte, North Carolina.

2005
Image Acts, Amnesty International Project, Office Ops, Brooklyn.
Workers of the World, Shenere Velt Gallery, Los Angeles.
Art / Work, Tufts University Gallery, Medford, Massachusetts.
The Simnuke Project, Rx Gallery, San Francisco.

2004
News from Home, The Annex, New York.
Felix Variations: Artists Respond to Felix Gonzalez-Torres, California College for the Arts, San Francisco.
Active Duty, Armed Artists of America, Studio 84, Brooklyn.
The Art of Politics, Ashmore Gallery, Miami.

2003
Les Fables de la Fontaine, The Jacob Lawrence Gallery, University of Washington, Seattle; The Meyerhoff Gallery, Baltimore; Exhibition Space of Temple University, Rome; Centre pour l'Art et la Culture, Institut Americain Universitaire, Aix-en-Provence.
The New Normal, Borofsky Gallery, Philadelphia.
Violent Violence, Arti et Amicitiae, Amsterdam.

2002
Art Basel / Miami, Art Positions, Miami.
Toxic Landscapes: Artists Examine the Environment, Reino de Este Mundo International Gallery, Biblioteca Nacional Jose Marti, Havana.

2001
Paradise in Search of a Future, CEPA, Buffalo, New York.

2000

Flesh and Blood V, John Batten Gallery, Hong Kong; 1708 Gallery, Richmond, Virginia; New Gallery, Miami; Lump Gallery, Raleigh, North Carolina.

1999

Requiem, Nexus Contemporary Art Center, Atlanta, Georgia.

1998

Context, Nexus Foundation for Today's Art, Philadelphia.

1997

Flesh and Blood, Hewlett Gallery, Carnegie Mellon, Pittsburgh.

1996

Female Trouble, City College Gallery, New York.

Revenge, site-specific installation, Chamber Gallery, New York.

1992

Neurotic Art: The Return of the Repressed, Artists Space, New York.

1991

Furious Women, Furious Work, Artemesia Gallery, Chicago.

Bibliography

Selected Publications by elin o'Hara slavick

2007

Bomb after Bomb. Milan: Edizioni Charta.

2006

"Workers Dreaming Photographs by elin o'Hara slavick," *Daylight Magazine*, no. 5, pp. 54–59.

Constantine and Koren Christofides, *Fables of La Fontaine*. Seattle–London: University of Washington Press, p. 68. The artist contributed a gouache painting on wood that illustrates one of the fables, "The Animals Sick from the Plague," Book VII, Fable 1.

"Field Report – Protesting Cartography: Places the U.S. Has Bombed," *Cultural Politics*, July, pp. 245–254.

2005

"Scraps of Hope for a Nameless Flag," in *Not in My Name: Silvia Velez*. Adelaide, Australia: Goanna Print, pp. 36–39.

2002

"Interview with Sue Coe," *The Zine Yearbook*, vol. 6, an Annual Collection of Excerpts from the Best Zines Publishing Today, Become the Media, pp. 87–91.

"Dissent from the Homeland: Essays after September 11" (photographs), *The South Atlantic Quarterly*, spring, p. 289.

2001

"Global Economy Postcard," *Adbusters*, January, p. 6.

Catherine Lutz, with photographs by elin o'Hara slavick, *Homefront: A Military City and the American Twentieth Century*. Boston: Beacon Press.

2000

"A Doll for the Global Economy," *The Progressive*, January, p. 15.

1994

"A Wretch like Me," edited by Robert Blanchon, *Whitewalls*, fall, p. 39.

Selected Publications about elin o'Hara slavick

2006

John Armitage, "Virilio over Hypermodern America: On the Recent Art of Jordan Crandall, Joy Garnett, and Elin O'Hara Slavick," in *Paul Virilio and the Arts Symposium*. Karlsruhe: ZKM.

2004

Ann Millett, "Loom 3: Labeler," *Art Papers*, September, pp. 44–45.

2003

Rutger Pontzen, "Critical Art from America," *Volkskrant*, July 19.

2002

Elizabeth Howie, "Protest Maps; slavick's Work Maps the Fatal Seduction of Distance," *Independent*, November 13; "Triangle Art Awards, elin o'Hara slavick, Artist / Activist," *Independent*, June 25.

2001

Clancy Nolan, "The Personal is Political," *Independent*, June 20, pp. 28–29.

2000

Todd Smith, Curator of American Art, Mint Museum, catalogue essay, in *New Frontiers IV*.

Dinah Ryan, "Flesh & Blood," *Art Papers,* September, pp. 43–44.

Tom Patterson, "Pictures with Punch," *Winston-Salem Journal*, April 30, p. E5; "Art as Politics," *Winston-Salem Journal*, April 9, pp. E1–E2.

1999

Carol Mavor, *Becoming*. Durham, North Carolina: Duke University Press, pp. 135–173.

1998

Carol Mavor, "Too Close to See," in *Flesh & Blood*, exhibition catalogue essay.

Linda Dougherty, "Studio Visit with elin o'Hara slavick," *Art Papers*, March, pp. 29–31.

Linda Dougherty, "Labor < > Leisure (Global Economy)," *Art Papers*, January, p. 59.

1997

Carol Mavor, "Collecting Loss," *Cultural Studies*, spring, pp. 113–137.

1995

Woody Holliman, "The Pleasures of Gender," *Art Papers*, July, pp. 58–59.

1994

Margaret Morgan, "From Dada to Mama: Feminism, the Ready-made and Contemporary Practice," *Binocular*, pp. 11–29.

1992

Jerry Tallmer, "The Dress off Her Back," *New York Post*, December 11.

Sources

Appy, Christian G., *Patriots: The Vietnam War Remembered from All Sides*. New York: Penguin Books, 2003.

Blum, William, *Killing Hope: U.S. Military and CIA Interventions since World War II*. Monroe, Maine: Common Courage Press, 1995; *Rogue State: A Guide to the World's only Superpower*. Monroe, Maine: Common Courage Press, 2000.

Bradley, James, *Flags of our Fathers*. New York: Bantam Books, 2000.

Castro, Raul, *Grenada: The World against the Crime*. Havana: Editorial de Ciencias Sociales, 1983.

Center for Land Use Interpretation, www.clui.org.

Ferguson, James, *Grenada: Revolution in Reverse*. New York: Monthly Review Press, 1990.

Freemantle, Tony, "The Price of Peace: The World Struggles to Purge the Poison after a Cold War Chemical Weapons Binge," *The Houston Chronicle*, 1997.

French, Howard, "100,000 People Perished, But Who Remembers?," *New York Times*, March 14, 2002; "Dark Side of U.S. Quest for Security: Squalor on an Atoll," *New York Times*, June 11, 2001.

Gallagher, Carol, *American Ground Zero: The Secret Nuclear War*. Cambridge, Massachusetts: MIT Press, 1993.

Gibb, Tom, *BBC News*, March 24, 2002.

Grossman, Zoltan, "A Century of U.S. Military Interventions: From Wounded Knee to Afghanistan," *Znet*, zmag.org, 2001.

Hacker, Barton C., *Elements of Controversy: The Atomic Energy Commission and Radiation Safety in Nuclear Weapons Testing, 1947–1974*. Berkeley: University of California Press, 1994.

Hales, Peter, *Atomic Spaces: Living on the Manhattan Project*. Chicago: University of Illinois Press, 1999.

Hersch, Seymour, "Watching Lebanon: Washington's Interest in Israel's War," *The New Yorker*, August 21, 2006.

Independent Commission of Inquiry on the U.S. Invasion of Panama, *The U.S. Invasion of Panama: The Truth Behind Operation "Just Cause."* Boston: South End Press, 1991.

Kirsch, Scott, *Proving Grounds: Project Plowshare and the Unrealized Dream of Nuclear Earthmoving*. New Brunswick, New Jersey–London: Rutgers University Press, 2005; "Peaceful Nuclear Explosions and the Geography of Scientific Authority," *Professional Geographer*, 2000.

Lindqvist, Sven, *A History of Bombing*. London: Granta Books, 2001.

Maclear, Kyo, *Beclouded Visions: Hiroshima-Nagasaki and the Art of Witness*. New York: State University of New York Press, 1998.

Makhijani, Arjun, Hu, Howard, and Yih, Katherine, *Nuclear Wastelands: A Global Guide to Nuclear Weapons Production and its Health and Environmental Effects*, by a Special Commission of International Physicians for the Prevention of Nuclear War and the Institute for Energy and Environmental Research. Cambridge, Massachusetts: MIT Press, 1995.

Masri, Rania, *Memories of the War*, unpublished speech, 2007.

Michaels, Anne, *Fugitive Pieces*. New York: A.A. Knopf, 1997.

Orion Magazine, November–December, 2006.

Robbins, Anthony, *Radioactive Heaven and Earth: The Health and Environmental Effects of Nuclear Weapons Testing in, on, and above the Earth*. New York: Apex Press, 1991.

Sebald, W.G., *On the Natural History of Destruction*. New York: Modern Library, Random House, 2004.

Shapiro, Michael J., *Violent Cartographies: Mapping Cultures of War*. Minneapolis–London: University of Minnesota Press, 1997.

Solnit, Rebecca, *Savage Dreams: A Journey into the Landscape Wars of the American West*. New York: Vintage, 1994.

St. Clair, Jeffrey, *In These Times*, 1999.

Trumbo, Dalton, *Johnny Got His Gun*. New York: Bantam Books, 1939.

U.S. Department of Energy, *Closing the Circle on the Splitting of the Atom: The Environmental Legacy of Nuclear Weapons Production in the United States and What the Department of Energy Is Doing about It*. Washington, D.C.: Office of Environmental Management, 1995.

Vanderbilt, Tom, *Survival City: Adventures among the Ruins of Atomic America*. New York: Princeton Architectural Press, 2002.

Virilio, Paul and Sylvere Lotringer, *The Accident of Art*. Cambridge, Massachusetts: Semiotext(e), MIT Press, 2005; *Art & Fear*. London–New York: Continuum, 2003; and Sylvere Lotringer, *Pure War*. Cambridge, Massachusetts: Semiotext(e), MIT Press, 1998.

The Washington Post, Friday, March 26, 1999.

Wasserman, Harvey and Solomon, Norman, *Killing Our Own: The Disaster of America's Experience with Atomic Radiation*. New York: A Delta Book, Dell Publishing, 1982.

Williams, Terry Tempest, *Refuge: An Unnatural History of Family and Place*. New York: First Vintage Books Edition, 1991.

Yamahata, Yosuke, *Nagasaki Journey: The Photographs of Yosuke Yamahata August 10, 1945*. San Francisco: Pomegranate Artbooks, 1995.

Zinn, Howard, *A People's History of the United States*. New York: HarperCollins, 1980; *Hiroshima: Breaking the Silence*. Westfield, New Jersey: Open Magazine Pamphlet Series, 1982; *The Zinn Reader: Writings on Disobedience and Democracy*. New York: Seven Stories Press, 1997; *Just War*. Milan: Edizioni Charta, 2006; *A Power Governments Cannot Suppress*. San Francisco: City Lights Books, 2007.

To find out more about Charta,
and to learn about our most recent
publications, visit

www.chartaartbooks.it

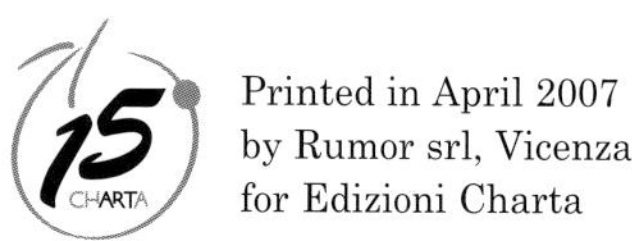

Printed in April 2007
by Rumor srl, Vicenza
for Edizioni Charta